Prof. Bharat Raj Singh,
Late (Prof.) Dinesh Jha

The Vibration of Heat Exchanger Tubes

Analysis & Design of Multiple Tubes
Vibration and Failure of Nuclear Power Plant

Lulu Press, Inc., USA

The Vibration of Heat Exchanger Tubes
Published on: Dec 07, 2019
First Edition

Master of Engineering (Analysis & Design of Process Equipments),
Department of Applied Mechanic.
Motilal Nehru Regional Engineering College, Allahabad.
Thesis approved by:
University of Allahabad, Allahabad, India) in July 1987.

ISBN: 978-1-79479-184-8

ID:	25855126
Category:	Engineering
Description:	The vibration of Heat Exchange Tubes due to hydrodynamic fluid coupling is an international problem for Nuclear fuel assemblies etc. on account of frequent failure of Heat exchanger tube, which causes not only expensive repair but a great loss to the plant. Thus, several studies in this field have been made so far. But here, a study of three circular cylindrical tubes in a liquid is done on the analytical approach. The author also describes the various parameters for maximum efficiency of heat transmission from Heat Exchangers.
Copyright Year:	2019
Language:	English
Country:	United States
Keywords:	Vibration, Fatigue Failure, Heat Exchanger,
Explicit Content:	No

Pricing & License
Price: US Dollar $10.49 (Rs.750/=)
License: Standard Copyright

Publisher:
Lulu Press, Inc.
627 Davis Drive, Suite 300, Morrisville, NC 27560,USA
www.lulu.com

Copyright © Lulu Press, Inc. All Rights Reserved.

Dedicated to

All mighty GOD and thereafter to my Father-in-Law;

Late Shri Dal Shringar Singh, MA (Economics), MA (English) and LT,

Ex-Principal, National Inter College, Kadipur, Sultanpur, India,

who was not only generous personality, but an inspiring and motivating factor

for my achievements.

Certificate

This is to certify that **Shri Bharat Raj Singh** worked under my supervision during the session 1986-87. He is submitting a thesis entitled "**Vibration of Heat Exchanger Tubes**" in partial fulfilment of requirement for the degree of Master of Engineering (Analysis and Design of Process Equipments) from the **University of Allahabad, Allahabad (U.P.)**. This work or part of it was not submitted anywhere else to obtain any degree or diploma.

                                          Dr. Dinesh Jha
                                          B.E. (Mechanical), M.E., Ph.D.
                                          Professor, Department of Applied Mechanics,
                                          Motilal Nehru Regional Engineering College,
                                          Allahabad-211004

Acknowledgement

The author is deeply indebted to **Dr. Dinesh Jha, Professor, Department of Applied Mechanics, Hydraulics and Hydraulic Machinery,** Motilal Neharu Regional Engineering College, Allahabad for his able guidance, kind help and encouragement in carrying out the present work. The author feels honoured to have worked under an eminent Professor like him in the field of Hydro-elastic Vibrations and Flow Induced Vibrations.

The author is thankful to **Dr. C. R. Rao, Professor and Head, Department of Applied Mechanics,** for encouragement from time to time. The author also likes to express his gratitude towards the teachers of the Applied Mechanics Department who have always extended helping hands whenever required.

The author also extends his gratitude to the library faculty, librarian and staff of Motilal Neharu Regional Engineering College (now MNNIT), Allahabad for providing the cooperation in various stages for preparation of the thesis work in 1987.

The author cannot forget to mention the support of his wife Smt. Malti Singh, children- Nidhi Singh, Saurabh Singh and Gaurav Singh who have always been motivating factor in shaping the thesis for the benefits of research scholar, academicians and industrialist.

Bharat Raj Singh
Allahabad, India.

Summary

Vibration of Heat exchange tubes due to hydrodynamic fluid coupling is an international problem for nuclear fuel assemblies etc. on account of frequent failure of Heat exchanger tube, causes not only expensive repair, but great loss to the plant. Thus, several studies in this field have been made so far. But here, study of three circular cylindrical tubes in a liquid is done on analytical approach. The author also describes the various parameters for maximum efficiency of heat transmission from Heat Exchanger's, which is defined as under, $\eta_H = F\left(\frac{G}{R}, V,\right)$ = Heat transmission efficiency of heat exchanger, G=gap between two adjoin tubes, R= Radius of cylindrical tubes (if considered of same diameter) and V= fluid flow velocity and geometry of tubes. The relation amongs above parameters are yet to derive to solve this problem.

Contents

S. No	Descriptions	Page No.
-	Dedication	(i)-(ii)
-	Certificate	(iii)-(iv)
-	Acknowledgements	(v)-(vi)
-	Summary	(vii)-(viii)
-	List of Figures	(xiii)
-	List of Tables, Appendices & Graphs	(xiv)-(xvi)
Chapter-I	Introduction	1-6
Chapter-II	Review of past works	7-20
Chapter-III	Fundamental of Vibrations	21-38
	3.1 Vibrations	23
	3.2 Element of Vibratory System	23
	3.3 Harmonic Vibration	24
	3.3.1 Simple Harmonic Motion	24
	3.3.2 Definitions	26
	3.3.3 Fourier Series	27
	3.4 Single Degree of Freedom Systems	29
	3.4.1 Free Vibrations without damping	29
	3.4.2 Damped Free Vibrations	31
	3.5 Transverse Vibrations of Beams	35

Chapter-IV	Dynamic Response of Two parallel Cylinders in a Liquid	39-54
	4.1 Equation of Motion	41
	4.2 Free Vibration	44
	4.2.1 Exact Solutions	44
	4.3 Approximate solution	48
	4.4 Forced Vibration	52
Chapter-V	Vibrations of Cylinders in a Liquid Having Square and other geometric Shapes	55-66
	5.1 Basic Equations	57
	5.1.1 Equations of continuity for square cylinders	57
	5.1.2 Equation of fluid Motion	60
	5.1.3 Equation of cylinder Motion	61
	5.2 Natural Frequencies	62
	5.2.1 Equation of Energy (Square Cylinders)	62
	5.2.2 Energy equation for Hexagonal Cylinder	64
	5.3 Frequency Response	65
Chapter-VI	Dynamic Response of Three parallel Circular Cylinders in a Liquid	67-82
	6.1 Dynamic Response of Three parallel Circular Cylindrical Tubes	69
	6.1.1 Nomenclatures	71

		6.1.2 Assumptions	72
	6.2	Free Vibration	75

Chapter-VII	Results and Discussions	83-90
	7.1 Parallel circular cylindrical rods immersed in liquid	85
	7.2 Square cylinders immersed in a liquid	87
	7.3 Three parallel circular cylindrical Tubes Immersed in a liquid	89

Chapter-VIII	Conclusions	91-94

Chapter-IX	References & Bibliography	95-100

Chapter-X	Tables, Appendices and Graphs	101-114
	Table-1 to Table-5	103-107
	Appendices-1 to 2	108-109
	Graphs Sheets No.-1 to 5	110-114

List of Figures

Fig. Nos.	Captions	Page Nos.
Figure 1.1	Schematic Diagram for Nuclear Power Reactor	4
Figure 1.2	Auxiliaries of Nuclear Power Reactor	4
Figure 3.0	Elements of Vibratory System	23
Figure 3.1	The Simple Vibrational Motion	24
Figure 3.2	Displacement (x) and Factors; ($m\dot{x}$, kx) of Vibratory Motion	25
Figure 3.3	Free Vibrations without Damping-Plot of Equation (**3.20**)	25
Figure 3.4	Diagram of the Mass, Spring and Damper for Damped Vibration	31
Figure 3.5	Vibration Amplitudes of Critically damped, Over damped, Under damped and Un-damped System	32
Figure 3.6	The Transverse Vibration of a Prismatic Beam	35
Figure 4.1	Two Parallel Circular Cylindrical Rods Immersed in a Liquid	41
Figure 5.1	Two-Dimensional Model of Square Cylinders in a Liquid	58
Figure 6.1	Three Parallel Circular Cylindrical Tubes in a Liquid	70

List of Tables, Appendices & Graphs

List of Tables

Table Nos.	Captions	Page Nos.
Table -1	Natural frequency to Different Tube Diameters & Thickness	103
Table -2	Mass Coefficient versus Gap Ratios of Tubes	104
Table- 3	Elements of Vibratory System	105
Table- 4	The Simple Vibrational Motion	106
Table-5	Displacement (x) and Factors; ($m\dot{x}$, kx) of Vibratory Motion	107

List of Appendices

Appendices Nos.	Captions	Page Nos.
Appendix -1	**Sample calculations:** Frequency Response of a Tube Bundle Emerged in Water	108
Appendix -2	**Sample Calculations:** Mass Coefficient with Different Gap Ratios	109

List of Graph Sheets

Graph Sheets Nos.	Captions	Page Nos.
Graph -1	Natural frequency Versus Different Tube Diameters & Thickness	110
Graph -2	Mass Coefficient Versus Gap Ratios of Tubes	111
Graph- 3	Frequency Ratio Versus Mass Ratios	112
Graph- 4	Frequency Ratio Versus Mass /Gap Ratio	113
Graph-5	Frequency Ratio Versus Tube Diameters with Different Thicknesses	114

CHAPTER-I

INTRODUCTION

1.
Introduction

For the design of heat exchangers and reactors international components such as fuel assemblies, the study of the vibration response of tube bundles in a liquid of various types of excitations including earth quakes, fluid flows and acoustic noises is of great importance. The several studies have been made on the coupled-motion of multi-rods in a liquid. The hydrodynamic effects for coupled system such as flow excited vibrations are causing great damage on nuclear power stations where heat generated through reactor is converted into steam through heat exchangers only. To get maximum advantages of the system fluid flow velocity can be increased, but it has got its own limitations. As soon as flow velocity is increased beyond its certain limits, the heat exchangers tubes gets vibrated due to flow excitations and ultimately at resonance conditions or at higher frequency it causes damage to heat exchanger tubes bundle. This is not only expensive to repair, but hampers the production of Electricity and reactor has to shut-off. A schematic system for nuclear power reactor is shown in **Fig. 1.1** and **Fig. 1.2**. Thus the study of spacing of tubes, diameter selection of tubes, flow velocity of fluid are of great importance to obtain maximum efficiency of the system.

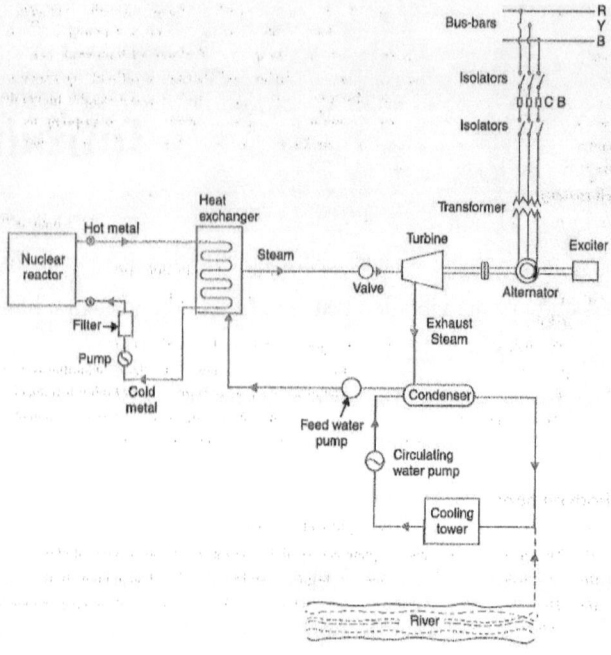

Figure 1.1: Schematic Diagram for Nuclear Power Reactor

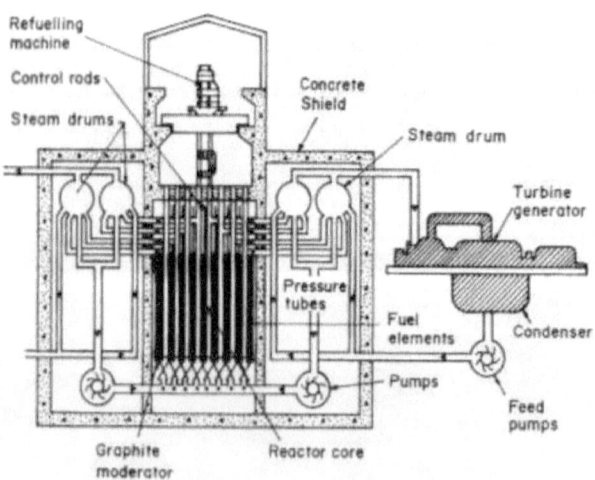

Figure 1.2: Auxiliaries of Nuclear Power Reactor

Lot of work is still continuing in this field and subject became international importance. Recently international conferences at Windermere in England on 12 to 14 May, 1987 was held on Flow Induced Vibrations, and under mentioned papers were presented specially for heat exchanger tube assembly vibrations.

a) Flow induced Acoustic Resonances in Heat Exchangers by G.J. Rae & B. G. Murray [**13**].
b) Two phase Buffeting of Heat Exchanger Tubes by H.G.D. Goyader [**14**].
c) Vibration of single Row of circular cylinders subjected to two phase Bubble Cross Flow [**15**].
d) Flow around multiple cylinders By Vortex Method by P.A. Smith. R. Renoyre [**16**].
e) A review of Flow Induced Vibrations in the Heat Exchangers by D.S. Weaver & J.A. Fitzpatrich [**17**].

Moreover, research work is still being carried out to estimate the effect of antivibration devices based on above work to avoid frequent and very expensive repairs of heat Exchangers. The study of tube-bundle vibration is done under various boundary conditions for better approach to the problem.

Looking into the importance of this topic, the author decided to work on the response of the system containing Tube Bundles, its frequency response characteristics and many other characteristics to be investigated.

Firstly, the review of past works is done (**Chapter-II**). Secondly the general study of vibrations single degree & beams is done (**Chapter-III**), the special study on Dynamic Responses of two parallel circular cylinders in a liquid is done (**Chapter-IV**) and on mathematical model as was proposed to describe the Dynamic behaviour of square & hexagonal cylinder Bundle immersed in a liquid (**Chapter-V**).

Lastly, the study of tube bundle based on three tubes dynamic responses is being undertaken. The efforts are also being made to generalize the case with tube-bundle, with mathematical approach and behaviour of fluid flow velocity, frequency under various parameters are to be studied for obtaining the maximum efficiency of the system (**Chapter-VI**).

CHAPTER-II

LITERATURE REVIEW

2.
Review of Past Work

In 1960, Masami Masubuchi [1] worked on Dynamic Response & Control of Multipass Heat Exchangers. The transfer functions obtained by dynamic analysis of one shell pass and 2, 3, 4.................2n, 2n+1 tube pass heat exchangers as a distributed parameter system are presented in dimensionless forms.

The heat exchange processes are found to be governed by the third-order characteristic equations with complex coefficients and can be solved numerically, using a graphical method. The numerical examples are presented to show the clear difference of frequency response for such cases when solved capacity exists. The analogy computer and the experimental results are found to be in good agreement with the theoretical results.

In 1960, Herman Thai Larsen [2] worked on Dynamics of Heat Exchangers and their models. The author rejects momentarily actual experimentation with real heat exchanger for logical cogitation and arm chair experiments. Observed as mental images or models these heat exchangers reveal their basic characteristics – characteristics that will be practically obscured in an actual experiment by secondary effects inherent in the equipment.

These basic characteristics once perceived may be used to complete by means of frequency response diagrams many results therefore diverse and seemingly unrelated that have been reported in the literature. Included among such results are not only the dynamic of various types of heat exchangers but also the thermal dynamic interaction between fluids and continuing pipe used for their transport. In addition, the author warns that all too frequently the use of the averaged temperature in heat exchanger dynamic is based on faulty logic.

In 1968 A. Protos, V.W. Goldschmidt and G.H. Toebes [3] worked on Hydro-elastic forces on Bluff cylinders. A bluff cylinder placed in flow will, in general, create an unsteady wake. As a consequence, it will be subject to the unsteady dynamic drag & lift forces.

In particular, it presents experimental findings that contribute both to the understanding of fluid elastic phenomena associated with periodic vortex formation as well as to the meagre found of accurate data useful in design.

In general, the resonant condition of a one degree of freedom system is characterised by the motions that are nearly harmonic and the study herein was made using cylinders in forced harmonic oscillation. The paper was concerned with flow induced resonance of bluff cylinder.

In 1973, P. W. Bearman and A. J. Wadcock [4] presented the paper on interaction between a pair of circular cylinders normal to a stream. This paper describes how the flows around two cylinders, displaced in a plane normal to the stream, interact as the two bodies are brought close together. Surface pressure measurements at a Reynolds number of 25×10^4 based on the diameter of a single cylinder; show the presence of mean repulsive force between the cylinders. Instability of the flow was fond when the gap between the cylinders was in the range between one diameter and about 0.1 of a diameter. Co-relation measurements of hot wire outputs indicate how mutual interference influences the formation of vortex streets from the two cylinders. Span-wise correlation measurements show that the correlation length doubles as the cylinders are brought into contact. They found that for a cylinder gap of a four diameters or more there was practically no interference.

In 1975, Sghoel Sheng Chen [5] worked on Dynamic Responses of two Parallel circular cylinders in a liquid. The problem of two parallel circular cylinders vibrating in a liquid is studied analytically. First, the equations of motion including fluid coupling are derived using the added mass concept.

Then, a closed form solution and an approximate solution are obtained for free vibration. Finally the study state responses of two cylinders subjected to harmonic excitations are presented. The results of this study illustrate the significance of the interaction of two structures in a liquid.

In 1981, Y. Shinohara & T. Shimogo [6] worked on a study on vibrations of square and Hexagonal cylinders in a liquid.

A mathematical model is proposed to describe the dynamic behaviour of square and hexagonal cylinder bundles immersed in a liquid. First the hydrodynamic forces associated with cylinder motion are examined, and then equations of motion of the spring mounted cylinders including liquid coupling are derived. When the number of cylinders are very large, these equations are replaced by partial differential equations on the assumption that the cylinder bundles form a continuum. The results of this study have application in the modelling of vibration of a nuclear fuel assembly under the excitation of earthquakes.

S.S. Chen & J.A. Jendrzejczyk [7] worked on Flow Velocity Dependence of Damping in Tube Arrays subjected to liquid cross Flow. Experiments are conducted to determine the damping for a tube in tube arrays subjected to liquid cross Flow; damping factors in the lift and drag directions are measured for in line and staggered arrays.

It is found that:

1) Fluid damping is not a constant, but a function of flow velocity.
2) Damping factors in the lift and drag directions are different.
3) Fluid damping depends on the tube location in an array.
4) Flow velocity dependent damping is coupled with vortex shedding process and fluid elastic instability and
5) Flow velocity dependent damping may be negative.

This study demonstrates that flow velocity dependent damping is important. These4 characteristic should be properly taken into account in the mathematics modelling of tube arrays subjected to cross flow.

In 1982, S. J. Brown [8] presented survey studies into the Hydrodynamic Response of Fluid coupled circular cylinder.

This survey concerns itself with the hydrodynamic analysis of fluid coupled co-axial cylinders under small displacements. Compressible fluids such as gases are not considered. The shell stiffness from rigid to flexible are considered as well as various combinations of fluid and structure boundary conditions. Cylindrical configurations may be found extensively used as pressure vessels, piping tubing, shafts, inserts, containers, barriers and structural members.

The hydrodynamic coupling of a fluid and structure is treated by many authors as consisting of two components of forces- Inertia & Damping. The inertia term has popularly been utilized as an added mass or virtual mass that is added to the actual mass of the structure.

The area covered by this survey is particularly important for several reasons:

1) A crucial step in the design of a plant, system apparatus, vessel or component operating in a fluid is sizing calculations based upon specification of operating emergency & faulted conditions. Material frequency analysis is always a preliminary to any dynamic analysis.
2) The effect of the fluid on the structures response is usually significant except for extremely thick vessels.
3) Usually no one analytical method is a panacea for a broad spectrum of boundary conditions and geometric complexities. An understanding of ranges of applicability is necessary.

In general, classical or simplified formed as have been found to be good design tools to perf9orm preliminary estimates of the response of fluid coupled cylinders for most nuclear PWR, fossil power, petrochemical, process and rotating equipment applications.

D. S. Weaver & D. Koroyannale is [9] worked on the cross-flow response of a tube array in water a comparison with same Array in Air. A water tunnel study was conducted on a parallel triangular array of tubes with a pitch ratio of 1.375. The array was geometrically identical to that used previously in a wind tunnel study so that the tube response to cross flow could be compared. It was seen that the response curves for tube arrays in water are much less regular than those in air, creating ambiguity in defining the stability threshold. The irregularities are seen to be associated with shifts in relative tube mode and frequency. Arrays in water apparently first become unstable in one of the lowest frequencies of the band of the frequencies associated with a given structural mode. The added mass coefficient computed from the observed frequency of instability is a little larger that the largest added mass coefficient obtained from the existing theory for tube array in quiescent fluid.

H. Tanaka, S. Takahara & K. Ohta [10] presented a paper on Flow-induced vibration of Tube Arrays with various pitch to Diameter Ratios. Tube arrays in cross flow start to vibrate abruptly when the flow reaches at a certain velocity. The threshold flow velocity depends upon the geometrical arrangement of tubes.

It is very important for practical applications to understand the relations between the threshold velocity and pitch to diameter ratio of tube array. Unsteady fluid dynamic forces on a tube array with a pitch to diameter ratio of 2.0 were classified experimentally and the characteristics of the threshold velocity were revealed by calculating the velocity with the unsteady forces. By comparing the threshold velocities of tube arrays of pitch to diameter ratio 2.0 and 1.33, the characteristics of threshold velocity with respect to pitch to diameter ratio were clarified.

K. Ohta, K. Kagawa, H. Tanaka & S. Takahara [11] worked on the fluid elastic vibration of Tube Arrays using Model Analysis Technique. This presents a method to calculate the critical flow velocity of fluid elastic vibration of tube arrays in heat exchangers. The method is based upon the mosal analysis technique, which combines the fluid dynamic force caused by cross flow and the vibration characteristics of the complicated tube arrays to obtain its response. The analytical method enables us not only to take into account the vibration mode of tube array and non uniformity of velocity and density distribution of cross flow but also to estimate the effect of anti vibration devices such as spacer, connecting band, and soon. Numerical examples of constrained single tube array, multi-tube array in reversed flow and group of panels with spacers are described.

S. S. Chen, J. A. Jenderzejc Zyk & M. W. Wambsganss [12] worked on the Dynamics of Tubes in Fluid with tube Baffle Interaction. Three series of tests are performed to evaluate the effects of tube to tube-support-plate (TSP) clearance on tube dynamic characteristics and instability phenomena for tube arrays in cross flow. Test results show that for relatively large clearances, tubes may possess TSP inn active modes in which the tubes rattle inside some of the tube support plate holes, and that the material frequencies of TSP in active modes are lower than those of "TSP-active modes", in which support plates provide 'knife-edge tube support. Tube response characteristics associated with TSP-inactive modes are sensitive to tube to-TSP clearance, TSP thickness, excitation amplitude, tube alignment and fluid inside the clearance. In addition tube response in intrinsically nonlinear, with dominance of TSP-in active or TSP-active modes depending on the magnitudes of different system parameters. In general, such a system is difficult to model; only full scale test can provide all the necessary characteristics.

A tube array supported by TSP's with relatively large clearances may be subjected to dynamic instability in some of the TSP inactive modes; tube response characteristics and impact forces on TSP's for a tube row are studied in detail. Tube displacements associated with instability of a TSP-inactive mode are small; however, impacts of the tube against TSP's may result in significant damage in relatively short time.

G. J. Rae & B. G. Murray [13] presented the paper on the flow induced Acoustic Resonances in Heat Exchangers were studied. This study presents some of the results an investigation into flow induced acoustic resonances in in-line heat exchangers. It has been clear for some time that the current models and design procedures for the avoidance of such acoustic resonances are both inadequate and contradictory in their approach to the problem. This study has re-investigated the nature and source of these resonances with the aim of providing a better insight to the mechanism.

This investigation has shown that:
1) The acoustic resonance can be excited at a frequency well removed from that vortex shedding. The results also show evidences of both phenomena existing, simultaneously at different frequencies.
2) The acoustic resonances behaviour is consistent with that of a self exited system.
3) An alternative model of this phenomenon provides a better procedure for avoiding these resonances in closely packed banks.

On the basis of these findings and the findings from a similar investigation into vortex induced tube vibration, it is proposed that tube vibration and acoustic vibration are treated as different mechanisms. Hence the guidelines presented in this paper should be followed to avoid the occurrence of acoustic vibration.

H.G.D. Goyder [14] studied on "Two Phase Buffeting of Heat Exchangers" Tubes. A gas liquid two-phase flow may be buffeting the tubes within heat exchanger and cause vibration. An equation for calculating the tube to support force which causes most damage is developed. This equation is specially formulated to give emphasis to the important non-model part of this force. An experimental technique is also described for measuring parameters such as tube to support forces, excitation spectra and flow correlation length scales.

Fumio-Hara [15] worked on vibration of a Single Row of circular cylinders subjected to Two-Phase Bubble Cross flow. Interesting feature connected with the fluid elastic vibration of tube bundles in cross flows, have been reported in many papers. Typically, experiments were done for single-phase liquid or gas flows. This paper presents experimental result on the vibration of a single row of five circular cylinders, having a ratio of pitch to diameter of 1.33, subjected to a two phase, mixed water-air bubble cross-flow. Test cylinders 25mm. In diameter and 58 mm ling horizontally installed in a rectangular test-section 200 mm long, 60 mm wide, and 200 mm high.

The five cylinders vibrated in both lift and drag directions and their vibrational acceleration being measured for different combinations of void fraction and reduced velocity. By changing the damping ratio of the cylinder system, we could cover a mass damping parameter from 0.88 to 4.72. relationship of vibration magnitude to avoid fraction, and/or reduced velocity and frequency characteristics such as power spectral density were investigated to clarify the nature of two phase flow induced vibrations for the row of circular cylinders, compared with that of the vibrations in a single-phase cross flow.

Experimental results showed the following major features:
1) Thy hysteresis character of the vibration magnitude in a single phase water flow disappeared for a two-phase bubble flow.
2) Four types of vibrational effects caused by air bubbles were found to appear, depending on the reduced velocity, mass-damping parameter and void fraction, i.e. for a small reduced velocity where the cylinder system was stable, air bubbles excited vibration randomly; for a flow velocity slightly higher than the critical void fraction, air bubbles, however, destabilized the vibration at a flow velocity a bit smaller than the critical; and for a fully supercritical flow velocity, air bubbles influenced vibration very little.
3) The instability map of fluid elastic vibration for a two phase bubble flow, with respect to the reduced velocity and mass-damping parameter, was differed form that for single-phase water flows, showing that the critical reduced velocity was greatly influenced by the air bubble concentration in the flow. A comparison of our experimental date with that of others indicated that most data plotted was within the instability region theoretically obtained by chen for a water cross flow.

P. K. Slansby, P.A. Smith, R. Penoyre [16] presented a paper on the flow Around Multiple cylinders by Vortex Method. This provides an interim status on the use of the vortex method, using random walks and the vortex in-cell technique, to predict post critical flow around multiple cylinders. The Baldwin-Pomax turbulence model is incorporated in the boundary-layer regions and mean pressure distributions are compared for one cylinder and two cylinders in various orientations with wind-Tunnel experiment using trip wire certain effects are predicted reasonably while others are not.

D. S. Weaver & J. A. Fitzpatrick carried out a Review of Flow induced Vibrations in Heat Exchangers [17]. Flow induced vibrations are widely recognized as a major concern in the design of modern tube & shell heat exchangers. Tube failures caused by excessive vibrations are relatively common place and often very expensive to repair. While considerable progress has been made inn the development of predictive tools, many uncertainties still remain. This paper reviews our state of understanding of the flow excitation mechanisms and presents design guide-lines. Also discussed are the research needs in this field.

In 1968, J. P. Glesing worked on "non linear Interaction of Two Lifting Bodies in Arbitrary unsteady Motion [18]. A method is developed for the exact calculation of the two dimensional potential flow about two bodies, either the body shapes or their motions. Pressures, forces, moments, and vortex-wake shapes are determined by applying a surface singularity method step by step in time. Calculated results for varied of flow situations are presented.

In 1975, S. S. Chen studied on "Vibrations of a Row of circular cylinders in a liquid [19]. The effects of interaction with surrounding liquid on the dynamic behaviour of row of circular cylinders are studied analytically. First the hydrodynamic forces associated with cylinder motions are obtained using the potential flow theory. Then a method of solution is presented for free and forced vibrations of a row of cylinders. The results of this study have important application in modelling of cross flow induced vibration of heat exchanger tubes.

In 1975, R. D. Blevins worked on Vibration of a Loosely Tube [20]. The results of a series of over 700 measurements of the tube damping and natural frequency of Heat Exchanger tubes are presented. The damping and frequency measurements of the loosely held tube show considerable scatter. Damping increased slightly with increased excitation force or decreasing pre load applied to the tube. Increasing the proportion of the tube length held in the support strongly increased damping, third mode vibrations were more lightly damped than first mode vibrations. The resonance frequency increased with static pre-load applied to the tube.

In 1977, B.T. Lubin, K.H. Halinger, a Puri and Gold berg studied, "The frequency Response of t Tube Bundle in water [21]. The first mode natural frequencies of tubes in a 3x5 array were found in air and water of primary interest were the natural frequency of two individually isolated tubes with all surrounding tube fixed, as compared with frequency response for the same pair of tubes with the surrounding tubes free. The first mode natural frequencies of these two isolated tubes are ranging from 123.6 to 129.3 Hertz. On the removal fixed condition, frequency spread over from 118 to 129.7 HZ. When system kept in water, the 15HZ reduction in frequency is caused by the hydrodynamic mass effect of the water and frequency ranged between 109.8 to 113.4 HZ.

In 1977, R. D. Blevins worked on Fluid Elastic Whirling of tube Rows and Tube Arrays [22]. Models are developed for the fluid force coefficients that determine the onset of whirling of tube rows and tube arrays. A control volume momentum analysis is employed. The results are in agreement with the available experimental date.

In 1977, S. S. Chen presented paper on Dynamics of Heat Exchanger tube Banks [23]. The paper presents that the flow induced vibration in heat exchangers used in nuclear reactor systems. In this paper the dynamic characteristics of tube banks in stationary liquid are studied. A method of analysis is presented for free and forced vibrations of tube banks including tube/ fluid interaction. Numerical results are given for tube banks subjected to various types of excitations.

In 1986, M.K. Auyang, worked on Turbulent Buffeting of a Multispan Tube Bundle [24]. An expression to calculate the buffeting response of Multispan Tube Bundle with non constant linear mass density is derived by generalizing Powell's joint acceptance concept. Application of the equation to lock in vortex induced vibration analysis is also discussed.

CHAPTER-III

FUNDAMENTALS OF VIBRATIONS

3.

Fundamentals of Vibrations

3.1 Vibration

The subject of vibration deals with the oscillatory motion of dynamic systems. All bodies possessing mass and elasticity are capable of vibration. Thus to and fro motion of a body about a mean position is called vibration.

3.2 Elements of vibratory system:

The elements that constitute a vibratory system are illustrated in **Fig. 3.0**. They are known as:

1) The mass (Inertia)
2) The spring (Elasticity)
3) The damper (fluid, viscosity or Atmospheric friction &)
4) The excitation (Disturbing force).

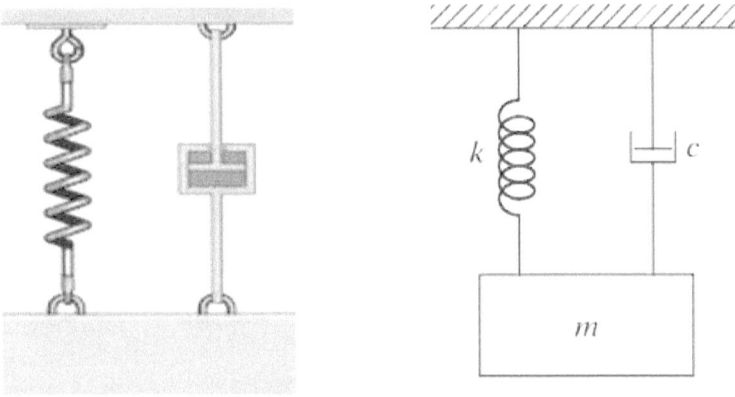

Figure 3.0: Elements of Vibratory System

3.3 Harmoning Vibration

3.3.1 Simple Harmonic Motion

Consider the simple equation of a Vibrational motion,

$$x = X \sin \omega t, \quad\quad\quad\quad\quad\quad\quad\quad\quad\quad\quad\quad\quad\quad\quad\quad (3.1)$$

when x is displacement and t is time.

Thus velocity of time 't'

$$\frac{dx}{dt} = \omega X \cos \omega t \quad\quad\quad\quad\quad\quad\quad\quad\quad\quad\quad\quad (3.2)$$

and acceleration at time 't'

$$\frac{d^2x}{dt^2} = -\omega^2 X \sin \omega t$$

$$= -\omega^2 z \quad\quad\quad\quad\quad\quad\quad\quad\quad\quad\quad\quad\quad\quad\quad (3.3)$$

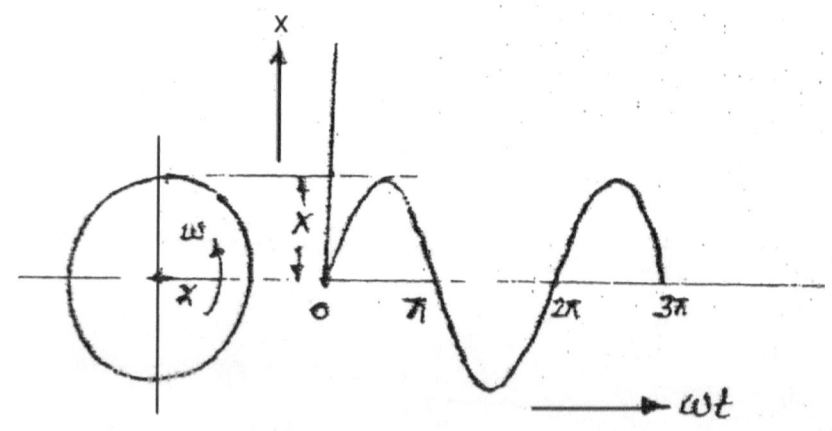

Figure 3.1: The Simple Vibrational Motion

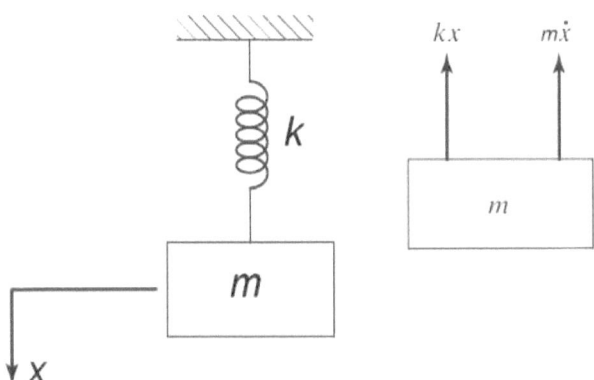

Figure 3.2: Displacement (*x*) and Factors; ($m\dot{x}$, kx) of Vibratory Motion

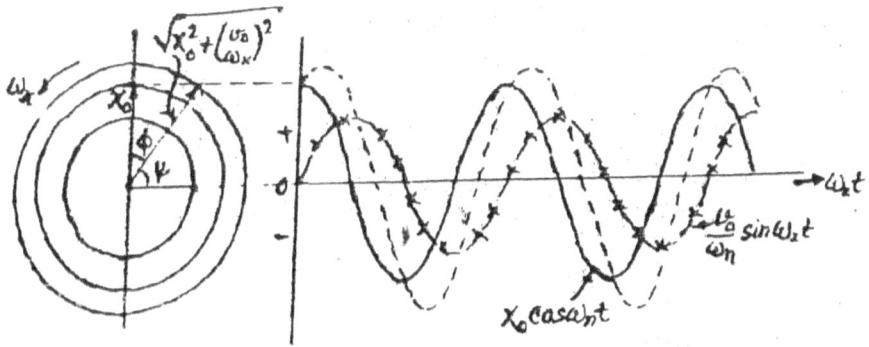

Figure 3.3: Free Vibrations without Damping-Plot of Equation (**3.20**)

Such a vibration where the acceleration is proportional to the displacement and is directed towards the mean position is a simple harmonic motion.

$$x = X \cos \omega t$$

is another example of a simple harmonic motion. This is also represented with a periodic function, and for completed vibration motion, it is analysed into its Fourier components.

If $x = X_1 \sin \omega t + X_2 \sin^2 \omega t + \dots$

$X_1 \sin \omega t$ is known as first harmonic

$X_2 \sin^2 \omega t$ is called as the second harmonic.

3.3.2 Definitions

If in equation
$x = X \sin \omega t$.

    X, ω are constants,

    X can be plotted against ωt (see **Fig. 3.1**)

Thus, vibrational displacement may be considered as the projection on the vertical axis of the end of a vector of length 'X' rotating at speed 'ω'.

Amplitude: Maximum displacement from the mean position of the vibrating body is called amplitude of its vibration.

Cycle: The movement of the body from the mean to its extreme position in one direction then to another extreme position and back to mean is called a "Cycle of Vibration".

Time period 'T' is the time taken to complete one cycle. It is equals the time for the vector to rotate through 2π and is therefore

$$T = \frac{2\pi}{w}$$

Frequency:

The number of cycles per unit time is called the frequency $= \frac{1}{T}$

No. of cycles per unit time = frequency

$$'F' = \frac{w}{2\pi} \quad \dots\dots\dots\dots\dots\dots\dots\dots\dots\dots(3.4)$$

Natural Frequency: If a system is disturbed and allowed to vibrate on its own, the frequency with which it vibrates without damping and without external forcing is known as its natural frequency and is denoted by 'ω_n'

the above definition of natural frequency is strictly applicable only to Single Degree of freedom system.

Phase Angle: If two vibrations are denoted by

$$x_1 = X_1 \sin \omega t \quad \quad (3.5)$$

$$x_2 = X_2 \sin (\omega t + \emptyset) \quad \quad (3.6)$$

Then second one leads the first one by a phase angle '$\emptyset$'

3.3.3 Fourier Series

A harmonic motion is simple and any periodic function of time $f(t)$ can be split up into an infinite series of terms forming the Fourier series

$$f(t) = \frac{a0}{2} + \sum_{n=1}^{\infty} (a_n \cos n\omega t + b_n \sin n\omega t) \quad \quad (3.7)$$

Where $\omega t = \frac{2}{T}$, T being the time period of $f(t)$

$$\text{and } a_n = \frac{w}{\pi} \int_{-\pi/w}^{+\pi/w} f(t) \cos n\omega t \, dt \quad \quad (3.8)$$

$$b_n = \frac{w}{\pi} \int_{-\pi/w}^{+\pi/w} f(t) \sin \omega t \, dt \quad \quad (3.9)$$

$$a_0 = \frac{w}{\pi} \int_{-\pi/w}^{+\pi/w} f(t) \, dt \quad \quad (3.10)$$

If $f(t)$ is an even function $a_n = 0$ & If $f(t)$ is an odd function $b_n = 0$.

Thus, the series can be represented by the sum of sines only as

$$F(t) = A_0 + A_1 \sin(\omega t + \varnothing_1) + A_2 \sin(2\omega t + \varnothing_2)$$

$$+ A_3 \sin(3\omega t + \varnothing_3) + \ldots \ldots \quad (3.11)$$

The second term is called fundamental or first harmonic of f (t) and the (n+1)th term of the series is called the nth harmonic comparing equation (3.11) & (3.7), we see that

$$A_n^2 = a_n^2 + b_n^2 \quad \ldots \ldots (3.12)$$

$$\tan \varnothing_n = \frac{a_n}{b_n} \quad \ldots \ldots (3.13)$$

Fourier series can be represented in terms of complex number because

$$\sin(n\omega t) = \frac{e^{J(n\omega t)} - e^{-J(n\omega t)}}{2j} \quad \ldots \ldots /\ldots (3.14)$$

$$\cos(n\omega t) = \frac{e^{J(\omega t)} + e^{-J(n\omega t)}}{2} \quad \ldots \ldots (3.15)$$

Thus, $$f(t) = \frac{a_o}{2} + \sum_{n=1}^{n=\infty} \left[a_n \left(\frac{e^{J(n\omega t)} + e^{-J(n\omega t)}}{2} \right) + b_n \left(\frac{e^{J(n\omega t)} + e^{-J(n\omega t)}}{2j} \right) \right]$$

$$\ldots \ldots (3.16)$$

$$= \frac{a_o}{2} + \sum_{n=1}^{\infty} \left[B.n.e.^{J(n\omega t)} + B(-n)e^{-J(n\omega t)} \right]$$

$$= \sum_{n=-\infty}^{n=+\infty} B_n e^{j(n\omega t)}$$

Where $B(n)$ and $B(-n)$ are complex conjugates

3.4 Single Degree of Freedom Systems

Single degree of freedom systems may be translational or torsional systems. The only condition is that it should be possible to describe the system configuration fully by single coordinate. For this, are four mathematical Techniques?

1. Energy method
2. Newton's Law of motion
3. Frequency response method &
4. Super portion Theorem

3.4.1 Free Vibrations without Damping (Translation System)

Consider the single degree of freedom spring mass system vibrating without damping and without forcing.

Let mass 'm' be given a displacement 'x', in the downward direction. The forces acting on the mass are:

(1) Inertia force $= - m \ddot{x}$ (in the upward direction0

(2) Spring force $= - k x$ (in the upward direction)

Now sum of the forces including inertia force on body in any direction must be zero as per 'D' Alembert's principle.

Therefore, $m \ddot{x} = 0$ .... (3.16)

Or $\ddot{x} + (\dfrac{K}{m}) x = 0$ .... (3.17)

Thus the solution of this differential equation would be

$X = A \sin \sqrt{\dfrac{k}{m}} t + B \cos \sqrt{\dfrac{k}{m}} .t$ .... (3.18)

Where A, B are constants to be determined from initial conditions.

The natural frequency ω_n of the system is given by

$$\omega_n = \sqrt{\frac{k}{m}} \text{ radian /Sec} \quad\quad\quad\quad\quad\quad\quad\quad\quad\quad\quad\quad\quad\quad\quad\quad (3.19)$$

Therefore linear frequency, $\omega = \dfrac{1}{2\pi}\sqrt{\dfrac{k}{m}}$ cycles cycles/sec.

If at t=0, $x = X_0$ and $\dfrac{dx}{dt} = v_o$

Substituting those conditions in the equation

$$A = \frac{V_o}{\omega_0}$$

$$B = X_0$$

So, $x = \dfrac{V_o}{\omega_0}\sin\left(\sqrt{\dfrac{K}{m}}\right) f + X_0 \cos\left(\sqrt{\dfrac{K}{m}}\right) t \quad\quad\quad\quad\quad\quad\quad (3.20)$

w_n may also be expressed in terms of the static deflection of the system. If w is the weight of mass 'm'

$$\begin{aligned}
w_n &= \sqrt{\frac{K}{m}} \\
&= \sqrt{\frac{K}{w/g}} \\
&= \sqrt{\frac{g}{w/k}} \\
&= \sqrt{\frac{g}{\delta}} \quad\quad\quad\quad\quad\quad\quad\quad\quad\quad\quad\quad\quad\quad\quad\quad (3.21)
\end{aligned}$$

Where δ is the static deflection. This expression applies only to the system if its configuration is vertical.

Equation (3.20) is graphically represented by **Fig. 3.3**.

It should be clear from the figure that the solution could be expressed as

$$x = \sqrt{x_0^2 + \frac{(v_0)^2}{w_n}} \cos(w_{nt} - \varnothing) \quad\quad\quad (3.22)$$

Where $\varnothing \tan^{-1}\left(\dfrac{v_0}{w_n x_0}\right)$

This can be rewritten as general solution,

$x = \sqrt{A^2 + B^2} \cos(w_{nt} - \varnothing)$

Where $\varnothing = \tan^{-1}\left(\dfrac{A}{B}\right)$

3.4.2 Damped Free Vibration

Introduce a damper to the system already studied and draw the free body diagram of the mass (**Fig. 3.4**).

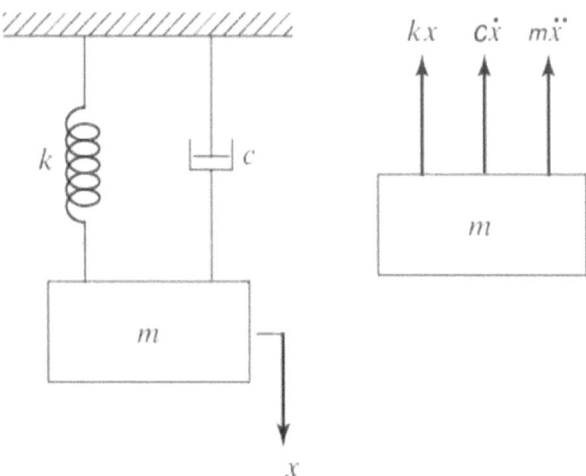

Figure 3.4: Diagram of the Mass, Spring and Damper for Damped Vibration

Summing up forces acting on the mass in the upward direction 'm'

$m\ddot{x} + c\dot{x} + kx = 0$

or $\ddot{x} + \left(\dfrac{c}{m}\right)\dot{x} + \left(\dfrac{k}{m}\right)x = 0$ ....(3.23)

Assume a solution in the form

$x = A_e{}^{pt}$ ....(3.24)

where A, p are constants to be determined

Differentiating the equation,

$\dfrac{dx}{dt} = \dot{x} = Ape^{pt}, \quad \dfrac{dx^2}{dt^2} = \ddot{x} = Ap^2e^{pt}$ ....(3.25)

Substituting (3.25) in equation (3.23), we obtain

$$\left\{p^2 + \left(\dfrac{c}{m}\right)p + \left(\dfrac{k}{m}\right)\right\}Ae^{pt} = 0$$

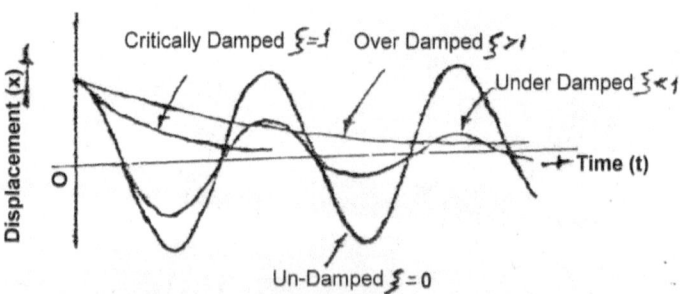

Figure 3.5: Vibration Amplitudes of Critically-damped, Over-damped, Under-damped and Un-damped System

or $P^2 + \left(\dfrac{c}{m}\right)P + \dfrac{K}{m} = 0$

from which two roots of P_1 & P_2 are given by

$$P_1 = \dfrac{c}{2m} + \sqrt{\left(\dfrac{c}{2m}\right)^2 - \dfrac{k}{m}}$$

and $\quad P_2 = \dfrac{-c}{2m} - \sqrt{\left(\dfrac{c}{2m}\right)^2 - \dfrac{k}{m}}$ ....(3.26)

Thus, $x = A e^{p_1 t} + B e^{p_2 t}$

Where A & B are constants whose values depend upon starting conditions.

The value of 'c' which makes the radical zero is called the critical damping is designated C_c.

Therefore, $\dfrac{C_c}{2m} = \sqrt{\left(\dfrac{k}{m}\right)} = \omega_n$

or $\quad C_c = 2m\,\omega_n$ ....(3.27)

The ration of $\dfrac{c}{C_c}$ is non dimensional and is called the damping ration ξ thus, c may be expressed as a

$\qquad c = \xi\, C_c$

and $\quad \dfrac{c}{2m} = \left(\dfrac{c}{C_c}\right)\left(\dfrac{C_c}{2m}\right) = \xi\,\omega_n$

Therefore, $P_1 = \left[-\xi + \sqrt{\xi^2 - 1}\right]\omega_n$

& $\qquad P_2 = \left[-\xi + \sqrt{\xi^2 - 1}\right]\omega_n$ ....(3.28)

The nature of roots P_1 and P_2 depends upon whether ξ is greater than, equal to or smaller than unity as shown in the **Fig. 3.5**.

(a) $\xi > 1$ over damping system:

P_1 & P_2 both are negative

So $x = Ae - (P_1)^t + Be - (P_2)^t$ ....(3.29)

Where (x) represents the magnitude

(b) $\xi < 1$ under damped system

$$P_{1,2} = \left[\xi \pm 1\sqrt{1 - \xi^2}\right]\omega_n \text{...(3.30)}$$

This shows that both the roots are complex

$$x = e^{-\xi w_n t}\left[Ae^{j\left(\sqrt{1-\xi^2}\right)W_n t} + B_e^{-j\left(\sqrt{1-\xi^2}\right)w_n t}\right]$$

$$= X_e^{-\xi w_n t}\cos\{(\sqrt{1-\xi^2})\omega_n t - \emptyset\} \text{...(3.31)}$$

This is a vibration with progressively diminishing amplitude. ω_d the frequency of vibration of the above system.

$$\omega_d = (\sqrt{1-\xi^2})\,\omega_n \text{...(3.32)}$$

(c) $\xi=1$ critically damped system:

It represents a case of transition between a periodic and oscillatory case. In this case P_1 and P_2 are the same and equal to $-W_0$. The solution may then be written as

$$x = (A + Bt)e^{-wnt}$$

3.5 Transverse Vibrations of Beams

Consider the transverse vibration of a prismatic beam is the x-y plane (**Fig. 3.6**), which is assumed to be a plane of symmetry for any cross section.

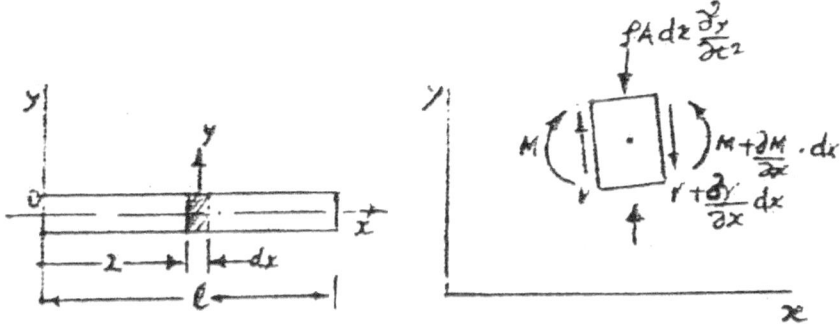

Figure 3.6: The Transverse Vibration of a Prismatic Beam

We shall use symbol 'y' to represent transverse displacement of a segment of the beam, located at the distance 'x' from the left hand end.

EL= flexural rigidity of the beam

dx= length of element

V= Shearing force

M= Bending Moment

When beam is vibrating transversely; the dynamic equilibrium condition for forces in the y direction is

$$V - \left(+\frac{V \partial v}{\partial y}.dx\right) - \rho.A.dx.\frac{\partial^2 y}{\partial t^2} = 0 \quad\quad\quad (3.31)$$

and moment equilibrium condition gives

$$-V\,dx + \frac{\partial M}{\partial x}dx \approx 0 \quad\quad\quad (3.32)$$

Substituting (V) from equ. (3.32) into (3.31).

$$\frac{\partial^2 M}{\partial x^2} dx = -\rho A \, dx \, \frac{\partial^2 y}{\partial x^2} \quad \text{...............(3.33)}$$

From elementary flexural theory, we have the relationship

$$M = EI \frac{\partial^2 y}{\partial x^2} \quad \text{...............(3.34)}$$

Using this expression in equation (3.33), we obtain

$$\frac{\partial^2}{\partial x^2}\left(EI \frac{\partial^2 y}{\partial x^2}\right) dx = \rho A \, dx \, \frac{\partial^2 y}{\partial t^2}$$

$$\text{or } EI \frac{\partial y}{\partial x^4} dx = -\rho A \, dx \, \frac{\partial^2 y}{\partial t^2}$$

This is general equation for transverse free vibration of a beam & can be re-written as

$$EI \frac{\partial^4 y}{\partial x^4} + (\rho A) \frac{\partial^2 y}{\partial t^2} = 0 \quad \text{...............(3.35)}$$

$$\text{or } \frac{\partial^4 y}{\partial x^4} + \frac{1}{a^2} \frac{\partial^2 y}{\partial t^2} = 0 \quad \text{...............(3.36)}$$

$$\text{where } a = \sqrt{\frac{EI}{\rho A}} \quad \text{...............(3.37)}$$

When a beam vibrates transversely in one of its natural modes, the deflection at any location varies harmonically with time, as follows

$$y = X (A \cos \omega t + B \sin \omega t) \quad \text{...............(3.38)}$$

Where the subscript 1 for the 1st mode has been omitted for notational convenience. Substituting equation (3.38) into equation (3.36) results in

$$\frac{d^4 x}{dx^4} - \frac{w^2}{a^2} x = 0 \quad \text{...............(3.39)}$$

Introducing $\frac{w^2}{a^2} = k^4$ ....(3.40)

Equation (3.39) can be re-written as,

$$\frac{d^4 x}{dx^4} - k^4 x = 0 \quad \text{...............(3.41)}$$

To satisfy equation (3.41), let

$$X = e^{px} \quad \text{...............(3.42)}$$

Thus, $\quad e^{PX}(p^4 - k^4) = 0$

Thus, the values of p are found to be p_1=k, p_2=-k, p_3=jk and p_4=-jk, where $j=\sqrt{-1}$
The general solution of equation (3.41) becomes

$$X = Ce^{kx} + De^{-kx} + Ee^{jkx} + Fe^{-jkx}$$

or $\quad X = C_1 \sin kx + C_2 \cos kx + C_3 \sinh kx + C_4 \cosh kx \quad \text{...............(3.43)}$

The expression represents a typical normal function for transverse vibrations of beam.

CHAPTER-IV

DYNAMIC RESPONSE OF TWO PARALLEL CYLINDERS IN A LIQUID

4.
Dynamic Response of Two Parallel Cylinders in a Liquid

The problem of two parallel circular cylindrical rods vibration was studied analytically by S.S. Chen [5] the effect of vibration response of rod bundles in liquids to various types of excitations (including earthquakes, fluid flows, and acoustic noises). First the equation of motion including fluid coupling are derived using the added mass coefficient. Then, a closed form solution and an approximate solution are obtained for free vibration; some important conclusion is drawn from the analysis. Finally steady state responses of two cylinders subjected to flow excitation are presented. The results of this study illustrate in significance of the interaction of two structures in liquid.

4.1 Equations of Motion

Consider two parallel circular cylindrical rods (**Fig. 4.1**) designated 1 & 2 immersed in a liquid. Rod motion consists of an emplane displacement along the y-axis and an out of plane displacement along the z-axis, can be written.

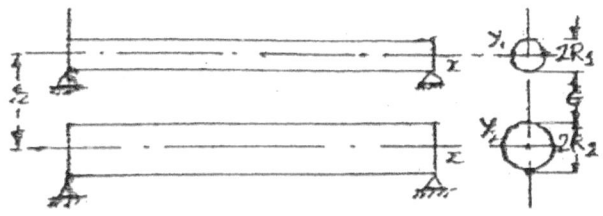

Figure 4.1: Two Parallel Circular Cylindrical Rods Immersed in a Liquid

$$E_j I_j \frac{\partial^4 u_j}{\partial x^4} + c_j \frac{\partial u_j}{\partial t} + m_j \frac{\partial^2 u_j}{\partial t^2} = F_j + f_j \dots \dots \dots \dots \dots (4.1)$$

Where J = denotes rod 1 (j=1) and rod 2 (j=2)
        X = axial coordinate
        t = time
        Uj = rod displacement
        Mj = mass per unit length of the rods
        EjIj = flexural rigidity
        Cj = damping coefficient
        Fj = hydrodynamic force
and   fj = excitation force

The hydrodynamic forces associated with two vibrating cylinders were considered by Mazur. Using a dimensional theory:

(a) For in plane motion

$$F_1 = -M_1/u_1 \frac{\partial^2 u_1}{\partial t^2} + M_1/u_3 \left(\frac{R_2}{r}\right)^2 \frac{\partial^2 u^2}{\partial t^2} \quad \&$$

$$F_2 = -M_2/u_2 \frac{\partial^2 u_2}{\partial t^2} + M_2/u_3 \left(\frac{R_1}{r}\right)^2 \frac{\partial^2 u^1}{\partial t^2} \dots \dots \dots \dots (4.2)$$

& (b) For out of plane motion

$$F_1 = -M_1 \mu_1 \frac{\partial^2 u_1}{\partial t^2} - M\mu_3 \left(\frac{R_2}{r}\right)^2 \frac{\partial^2 u^2}{\partial t^2}$$

& $$F_2 = -M_2 \mu_1 \frac{\partial^2 u_2}{\partial t^2} - M_2\mu_3 \left(\frac{R_1}{r}\right)^2 \frac{\partial^2 u_1}{\partial t^2} \dots \dots \dots \dots (4.3)$$

Where M_1 & M_2 are displaced masses of fluid by two rods R_1 & R_2 are radial of rods

r is distance between centre of two rods

μ_1, μ_2 & μ_3 are added mass coefficients

$$\mu_1 = 1 + \frac{r^4 - 2r^2(R_1^2 + R_2^2) + (R_2^2 + R_1^2)}{r^2 R_1^2} \sum_{K-1}^{\infty} k \frac{e^{-k(h+h_1)}}{\sin(kh)} \quad \&$$

$$\mu_2 = 1 + \frac{r^4 - 2r^2(R_1^2 + R_2^2) + (R_2^2 - R_1^2)}{r^2 R_2^2} \sum_{K-1}^{\infty} k \frac{e^{-k(h+h2)}}{\sin h(kh)} \quad \text{...(4.4)}$$

and $\mu_3 = 1 + \frac{r^4 - 2r^2(R_1^2 + R_2^2) + (R_2^2 - R_1^2)^2}{r^2 R_2^2} \sum_{K-1}^{\infty} h \coth(kh) e^{-2kh}$

where $h = \ln\left\{ \frac{r^2 - R_1^2 - R_2^2}{2R_1 R_2} + \left[\left(\frac{r^2 - R_1^2 - R_2^2}{2R_1 R_2} \right)^2 - 1 \right]^{\frac{1}{2}} \right\}$

$h = 2 \ln\left\{ \frac{r^2 + R_1^2 - R_2^2}{2rR_1} + \left[\left(\frac{r^2 + R_1^2 - R_2^2}{2rR_1} \right)^2 - 1 \right]^{\frac{1}{2}} \right\}$

and $h_2 = 2 \ln\left\{ \frac{r^2 - R_1^2 - R_2^2}{2rR_2} + \left[\left(\frac{r^2 - R_1^2 + R_2^2}{2rR_2} \right)^2 - 1 \right]^{\frac{1}{2}} \right\} \quad \text{...(4.5)}$

The values of $U_k = (k=1, 2, 3)$ depends on the dimensionless parameters $\frac{R2}{R1}$ & $\frac{G}{R_1}$

where $G = r - R_1 - R_2$ is gap between two rods.

Added mass coefficients $\rightarrow \frac{G}{R_1}$ when $\frac{R2}{R1} = 1.0$

These effects are studied at various conditions of $\frac{G}{R_1} = 0.5,$ and 1

Table: 4.2 shows the figures and graphs are plotted at various conditions **Fig. (4.3)**

Substituting equations (4.2) and (4.3) into (4.1), we find

$$E_1 I_1 \frac{\partial^4 4_1}{\partial x^4} + C_1 \frac{\partial u1}{\partial t} + (m1 + \mu_1 M_1) \frac{\partial^2 u1}{\partial t^2} - \lambda M_1 \mu \left(\frac{R2}{r} \right) \frac{\partial^2 u2}{\partial t 2} = f_1$$

$$E_2 I_2 \frac{\partial^4 u_2}{\partial x^4} + C_2 \frac{\partial u_2}{\partial t} + (m_2 + \mu M_1) \frac{\partial^2 u_2}{\partial t^2} - \lambda M_2 \mu_3 \left(\frac{R_1}{r} \right) \frac{\partial^2 u_1}{\partial t^2} = f_2$$

Wher $\lambda = 1$ for in plane motions

and $\lambda = -1$ for out of plane motions.

Rod motions in the two planes are uncoupled.

4.2 Free Vibration

4.2.1 Exact Solution: Neglect damping terms and forcing functions in equation (4.5) and let

$$U_j = R_j U_j e^{iwt} \quad\quad\quad\quad\quad\quad\quad\quad\quad\quad (4.6)$$

Where $1 = \sqrt{-1}$, w = Circular vibration frequency

w_{jn} = natural frequency of nth mode of the cylinder j in liquid.

R_j = cylinder radii

u_j = cylinder displacement

$f_j = 0$ & $C_j = 0$

u_j = Arbitrary constant and function of amplitude

Therefore when j=1,

$$u_1 = R_1 \overline{u}_1 \, Cos \, wt + 1 Sin \, wt) = R_1 \overline{U}_1 e^{iwt} \quad\quad\quad\quad (a)$$

& $u_2 = \overline{u}_2 \, R_2 \, e^{iwt}$

Therefore,

$$\frac{\partial u_1}{\partial t} = \overline{u}_1 \, R_1 I \, w e^{2wt}$$

$$\frac{\partial^2 u_1}{\partial t^2} = \overline{u}_1 \, R_1 (i)^2 w^2 e^{iwt} = -R_1 W^2 \, \overline{u}_1 e^{+iwt} \quad\quad\quad\quad (b)$$

Also $\dfrac{\partial u_1}{\partial x} = R_1 e^{+iwt} \dfrac{\partial \overline{u}_1}{\partial x}$

or $\dfrac{\partial^4 u_1}{\partial x^4} = R_1 \overline{e}^{iwt} \dfrac{\partial^4 \overline{u}_1}{\partial x_4}$ ....(c)

Similarly, $j = 2, u_{20} = R_2 \overline{u}_2 e^{iwt}$ ....(d)

$\dfrac{\partial u_2}{\partial t} = R_2 \overline{u}_2 \, iw e^{iwt}$

$\dfrac{\partial^2 u_2}{\partial t} = -R_2 \overline{u}_2 \, w^2 \, e^{iwt}$ ....(e)

Also, $\dfrac{\partial u_2}{\partial x} = R_2 e^{iwt} \dfrac{\partial \overline{u}_2}{\partial x}$ & $\dfrac{\partial^4 u_2}{\partial x^4} = R_2 e^{iwt} \dfrac{\partial^4 \overline{u}_2}{\partial x_4}$ ....(f)

Substituting, the values of $\dfrac{\partial^2 u_1}{\partial t^2}, \dfrac{\partial^2 u2}{\partial t^2}, \dfrac{\partial^4 u_1}{\partial x^4}, \dfrac{\partial^4 u_2}{\partial x^4}$

and neglecting damping factor $C_1 \text{ or } C_2 = 0$ & $f_1 = f_2 = 0$ in (4.5), we get

$$E_1 I_1 R_1 \dfrac{\partial^4 \overline{u}_1}{\partial x^4} - w^2 (m_1 + \mu_1 M_1) \overline{u}_1 R_1 + w^2 R_2 \lambda . M_1 \mu_3 \left(\dfrac{R_2}{r}\right)^2 \overline{u}_2 = 0 \;.$$

Or $\dfrac{\partial^4 \overline{u}_1}{\partial x^4} - \dfrac{m_1 + \mu_1 M_1}{E_1 I_1} w^2 \overline{u}_1 + \dfrac{\lambda M_1 \mu_3 w^2 \left(\dfrac{R2}{r}\right)}{E_1 I_1} \overline{u}_2 = 0$

Similarly $\dfrac{\partial^4 \overline{u}_2}{\partial x^4} - \dfrac{m_2 + \mu_2 M_1}{E_2 I_2} w^2 \overline{u}_2 + \dfrac{\lambda M_2 \mu_3 w^2 \left(\dfrac{R_1}{r}\right)^2}{E_{21} I_2} \overline{u}_1 = 0$ ....(4.7)

The solution of equation (4.7)

$$\overline{u}_1 = \sum_{n=1}^{8} a n \, e^{(pn^x)} \text{ and } \overline{u}_2 = \sum_{n=1}^{8} a n \, \gamma n \, e^{(pn^x)} \quad\quad\quad\quad (4.8)$$

Now, differentiating the above equations

$$\frac{\partial \bar{u}_1}{\partial x} = \sum_{n=1}^{8} anp_n \left[-\operatorname{Sin} P_n^{x} i \cos p_n^{x} \right]$$

$$\frac{\partial^2 \bar{u}_1}{\partial x^2} = \sum_{n=1}^{8} anp^2{}_n \left[-\operatorname{Cos}_n^{x} - i \operatorname{Sin} p_n^{x} \right]$$

$$\frac{\partial^3 \bar{u}_1}{\partial x^3} = \sum_{n=1}^{8} anp^3{}_n \left[-\operatorname{Sin}_n^{x} - i \operatorname{Cos} p_n^{x} \right]$$

$$\frac{\partial^4 \bar{u}_1}{\partial x^4} = \sum_{n=1}^{8} anp^4{}_n \left[-\operatorname{Cos}(P_n^{x}) + i \operatorname{Sin}(Pn^{x}) \right]$$

Therefore,

$$\frac{\partial^4 \bar{u}_1}{\partial x^4} = \sum_{n=1}^{8} anp^4{}_n \, e^{pn^x}$$

$$\frac{\partial^4 \bar{u}_2}{\partial x^4} = \sum_{n=1}^{8} anrn \, P^4{}_n \, e^{pn^x} \quad \dots\dots\dots\dots\dots\dots\dots\dots\dots\dots\dots\dots\dots\dots\dots\dots \quad (4.8.1)$$

From (4.8) $\bar{u}_1 = m \sum_{n=1}^{8} an \, e^{pn^x} = \gamma_n \bar{u}_1$

Thus, $\gamma_n = \dfrac{\bar{u}_2}{\bar{u}_1}$ ……………………………………………………………………(4.8.2)

Substituting (4.8.1) into (4.7), which gives

$$\sum_{n=1}^{8} a\, n\, p_n^4 e\, Pn^x - \frac{m_1 + \mu M_1}{E_1 I_1} w^2 \bar{u}_1 + \frac{\lambda \mu_3 M_1}{E_1 I_1} \cdot \left(\frac{R_2}{r}\right)^2 \cdot w^2 \bar{u}_2 = 0$$

Substituting (4.8.2) & (4.8) in above equation, it gives

$$p_n^4 \bar{u}_1 - \frac{m_1 + \mu M_1}{E_1 I_1} w^2 \bar{u}_1 + \frac{\lambda \mu_3 M_1}{E_1 I_1} \cdot \left(\frac{R_2}{r}\right)^2 \cdot w^2 \gamma \bar{u}_1 = 0$$

Vibration of Heat Exchanger Tubes 47

Or $\gamma = \dfrac{(m_1+\mu M_1)w^2 - E_1 I_1}{\lambda M_1 \mu_3 \left(\dfrac{R_2}{r}\right)^2 w^2} P_n^4 = \dfrac{\bar{u}_2}{\bar{u}_1}$ ....(4.9)

and, $\displaystyle\sum_{n=1}^{8} \gamma n\, p_n^4 e\, Pn^x - \dfrac{m_2 + \mu M_2}{E_2 I_2} w^2 \bar{u}_2 + \dfrac{\lambda \mu_3 M_2}{E_2 I_2}\left(\dfrac{R_1}{r}\right)^2 . w^2 \bar{u}_1 = 0$

or $P_n^4 e\bar{u}_2 - \dfrac{m_2 + \mu_2 M_2}{E_2 I_2} w^2 \bar{u}_2 + \dfrac{\lambda \mu_3 M_2}{E_2 I_2}\left(\dfrac{R1}{r}\right)^2 . w^2 \bar{u}_1 = 0$

or $\bar{u}_2 \left[P_n^4 - \left(\dfrac{m_2 + \mu_2 M_2}{E_2 I_2}\right) w^2 \right] + \dfrac{\lambda \mu_3 M_2}{E_2 I_2}\left(\dfrac{R_1}{r}\right)^2 . w^2 \bar{u}_1 = 0$

Substituting the value of $\bar{u}_2$ from equation (4.9), we get

$\dfrac{(m_1 + \mu M_1) w^2 - E_1 I_1 P_n^4}{\lambda M_1 \mu_3 \left(\dfrac{R}{r}\right)^2 w^2}\left[P_n^4 - \dfrac{(m_2 + \mu M_2)}{E_2 I_2} w^2 \right] + \dfrac{\gamma \mu_3 M_2}{E_2 I_2}\left(\dfrac{R_1}{r}\right)^2 W^2 = 0$

or $P_n^4\left[(m_1 + \mu M_1)w^2 - E_1 I_1 P_n^4\right] - \left[(m_1 + \mu_1 M_1)w^2 - E_1 I_1 P_n^4\right]$

$\left(\dfrac{m_2 + \mu_2 M_2}{E_2 I_2}\right) w^2 + \dfrac{M_1 M_2 \mu_3^2}{E_2 I_2}\left(\dfrac{R_1 R_2}{r}\right)^2 . w^4 = 0$

or $P_n^8 E_1 I_1 - P_n^4\left[(m_1 + \mu_1 M_1)w^2 + E_1 I_1 \left(\dfrac{m_2 + \mu_2 M_2}{E_2 I_2}\right) w^2\right]$

$+ w^4 \dfrac{(m_1 + \mu_1 M_1)(m_2 + \mu_2 M_2)}{E_2 I_2} - \dfrac{\lambda M_1 M_2 \mu_3^2}{E_2 I_2}\left(\dfrac{R_1 R_2}{r}\right)^2 . w^4 = 0$

Hence, $P_n^8 - w^2 P_n^4 \left(\dfrac{m_1 + \mu_1 M_1}{E_1 I_1} + \dfrac{m_2 + \mu_2 M_2}{E_2 I_2}\right)$

$$+\left[\frac{(m_1+\mu_1 M_1)}{E_1 I_1}\frac{(m_2+\mu_2 M_2)}{E_2 I_2}-\frac{\lambda\mu_3^2 M_1 M_2}{E_1 I_1 E_2 I_2}\left(\frac{R_1 R_2}{\pi}\right)^2\right]w^4=0 \quad\ldots\ldots\ldots\ldots(4.10)$$

Where from equation (4.9), $\quad \dfrac{\overline{u_2}}{u_2} = \dfrac{(m_1+\mu M_1)w^2 E_1 I_1 P_n^4}{\lambda M_1 \mu_3 \left(\dfrac{R2}{r}\right)^2 w^2}$

Substitution of equation (4.8) into boundary c conditions at x=0 and x=1 (l=is rod length), yields

$\{bmb\}\ \{an\}\quad =\quad \{0\}$ ....(4.11)

The element b_{mn} depend on frequency, fluid density f, material and geometric properties of rods and end conditions. Therefore, from equation (4.11) frequency equation may be written as

$F(w, E_j, E_j, m_j, R_j, 1, f, G) = 0$ ....(4.12)

4.3 Approximate Solution

It is of little trouble to obtain the frequencies from the exact frequency equation. However, it is desired to examine approximations to the frequency. First, consider a limiting case; one of the rod is rigid. The equation for free vibration in this case is

$$E_j I_j \frac{\partial^4 u_j}{\partial x^4} + (m_j + \mu_j M_j)\frac{\partial^2 u_j}{\partial t^2} = 0 \quad\ldots\ldots\ldots\ldots(4.13)$$

Let the natural frequency of the nth mode of the rod (j) in VACUO be denoted by, $\overline{w_{jn}}$. It is easily shown from (4.13) that the frequency for the rod close to a rigid rod is

$$w_{jn} = \frac{\overline{w_{jn}}}{\sqrt{1+\mu_j \beta_j}} \quad \text{where} \quad \beta = \frac{M_j}{m_j}$$

The corresponding modal functions satisfy the relation,

$$E_j I_j \frac{\partial^4 \emptyset_{jn}}{\partial x^4} - W_{jn}^2 (m_j + \mu_j M_j) \emptyset_{jn} = 0 \quad \text{...........(4.15)}$$

Thus, the natural frequencies are reduced in proportion to $\frac{1}{\sqrt{1+\mu_j \beta_j}}$ & modal functions are exactly same as those in vacuo

$$\text{Let } U_j = R_j \sum_{n=1}^{8} q_{jn}^{(t)} \emptyset_{jn}(x) \quad \text{...........(4.16)}$$

Where $\emptyset_{jn}(x)$ are orthonormal functions of rods in vacuous. Substituting equation (4.16) into (4.15) and using orthogonal conditions yield.

$$\ddot{q}_{1n} + 2\xi_{1n} w_{1n} \dot{q}_{1n} + w_{1n}^2 q_{1n} - \lambda \alpha_1 \Sigma_m \cdot a_{nm} \ddot{q}_{2m} = Q_{1n}(t)$$

$$\ddot{q}_{2n} + 2\xi_{2n} w_{2n} \dot{q}_{2n} + w_{2n}^2 q_{2n} - \lambda \alpha_2 \Sigma_m \cdot a_{mn} \ddot{q}_{1n} = Q_{2n}(t) \text{..........(4.17)}$$

where $\alpha_j = \frac{\beta_j \mu_3 (R-R-G)^3}{R_j R^2 (1+\mu_3 \beta_j)} \alpha_2$

$\xi_{in} = \frac{C_j}{2 w n_{jn}(m_j + \mu \ M_j)} \alpha_2$

$a_{nm} = \frac{1}{\ell} \int_0^\ell \emptyset_{in} \emptyset_{2m} dx,$

and $Q_{jn} = \frac{1}{1 R_j (m_j \mu, M_j)} \int_0^\ell f_j(x,t) \emptyset_{jn} dx, \quad \text{...........(4.18)}$

For free vibration, neglect the damping & forcing term

and let $Q_{jn} = \bar{q}_{jn} e^{iwt}$ ....(4.19)

For Rod I, J=1

$$q_{1n} = \overline{q_{1n}}(\cos \omega t + i \sin \omega t)$$
$$\dot{q}_{1n} = \overline{q_{1n}} \omega (-\sin \omega t + i \cos \omega t)$$

$$\ddot{q}_{1n} = -\bar{q}_{1n}\omega^2(\cos\omega t + i\sin\omega t) = -\bar{q}_{1n}\omega^2 e^{i\omega t}$$

For Rod 2, J=2

$$q_{2n} = \bar{q}_{2n}(\cos\omega t + i\sin\omega t)$$
$$\ddot{q}_{2n} = -\bar{q}_{2n}\omega^2 e^{i\omega t}$$

Substituting these in (4.17)

$$-\bar{q}_{1n}w_e^{2iwt} + w_{1n}^2(+\bar{q}_{1n}e^{iwt}) - \lambda\alpha_1\sum_m(-\bar{q}we_{2m}^{2\ iwt})anm = 0$$

$$-\bar{q}_{2n}w_e^{2iwt} + w_{2n}^2(+\bar{q}_{2n}e^{iwt}) - \lambda\alpha_2\sum_m(-\bar{q}_{1m}^{-2}e^{iwt})anm = 0$$

Or $(w^2 - w^2{}_{1n})\bar{q}_{1n} - \lambda\alpha_1 w^2 \sum_m anm\ \bar{q}_{2m} = 0$

$(w^2 - w^2{}_{2n})\bar{q}_{2n} - \lambda\alpha_2 w^2 \sum_m anm\ \bar{q}_{1m} = 0$ ....(4.20)

Equation (4.20) consists of an infinite number of ordinary equations. However, typically only finite number of equation is taken from case to case according to the desired accuracy. The frequency equation obtained from equation is

$$\begin{vmatrix} A & B \\ C & D \end{vmatrix} = 0 ....(4.21)$$

Where A, B, C, D are matrix element $A_{mn} = (w^2 - w_{1n})\delta mn$

$$B_{mn} = \lambda\alpha_1 w^2 a_{nm}$$
$$C_{mn} = -\lambda\alpha_2 w^2 a_{amn}$$
$$D_{mn} = (w^2 - w_{2n})\delta_{mn} ....(4.22)$$

When two rods have the same type of boundary condition

$\emptyset_{1n} = \emptyset_{2n}$ and $a_{mn} = \delta_{mn}$ ....(4.23)

Thus, equation (4.20) becomes,

$$\begin{vmatrix} w^2 - w_{1n}^2, -\lambda\alpha_1 w^2 \\ -\lambda\alpha_2 w^2, w^2 - w_{2n}^2 \end{vmatrix} \begin{Bmatrix} \overline{q}_{1n} \\ \overline{q}_{2n} \end{Bmatrix} = \begin{Bmatrix} 0 \\ 0 \end{Bmatrix} \quad \ldots(4.24)$$

Then frequency equation becomes,

$$(1-\alpha_1\alpha_2)w^4 (w_{1n}^2 + w_{2n}^2)w^2 + w_{1n}^2 w_{2n}^2 = 0 \quad \ldots(4.25)$$

Equation (25) gives two frequencies Ω_{1n} and Ω_{2n}

$$\Omega_{1n} \left\{ (w_{1n}^2 + w_{2n}^2) - \left[(w_{1n}^2 - w_{nn}^2)^2 + 4\alpha_1\alpha_2 w_{1n}^2 w_{2n}^2 \right]^{\frac{1}{2}} \right\}^{\frac{1}{2}}$$

$$\& \Omega_{2n} \left\{ (w_{1n}^2 + w_{2n}^2) - \left[(w_{1n}^2 - w_{2n}^2) + 4\alpha_1\alpha_2 w_{1n}^2 w_{2n}^2 \right]^{\frac{1}{2}} \right\} \left[2(1-\alpha_1\alpha_2) \right]^{\frac{1}{2}} \quad \ldots(4.26)$$

The amplitude ration, $\overline{q}_{2n}/\overline{q}_{1n}$ is given by

$$\frac{\overline{q}_{2n}}{\overline{q}_{1n}} = \frac{\Omega_{jn}^2 - w_{1n}^2}{\lambda\alpha_1\Omega_{jn}^2} \quad \ldots(4.27)$$

When $\alpha_1, \alpha_2 < 1$ ....(4.28)

Using in (4.25)

$$\Omega_{1n} < w_{1n}, w_{2n} \quad \& \quad \Omega_{2n} > w_{1n}, w_{2n} \quad \ldots(4.29)$$

When two tubes are identical $w_{1n} = w_{2n}$ & $\alpha_1 = \alpha_2$

$$\Omega_{1n} = \frac{w_{1n}}{\sqrt{1+\alpha 1}}, \frac{\overline{q}_{2n}}{\overline{q}_{1n}} = -\lambda$$

$$\& \Omega_{2n} = \frac{w_{1n}}{\sqrt{1+\alpha 1}}, \frac{\overline{q}_{2n}}{\overline{q}_{1n}} = \lambda \quad \ldots(4.30)$$

4.4 Forced Vibration

The steady state response of two rods is considered it is assumed that two simplify supported into al subjected to an excitation of the following form

$$f_1(x,t) = g_1(x)\sin \omega t.$$

and
$$f_2(x,t) = g_2(x) \sin(\omega t + \chi) \quad \ldots\ldots\ldots\ldots\ldots(4.31)$$

The response can be obtained from equations (4.16) and (4.17). In this case, equation (4.17) becomes:

$$\ddot{q}_{1n} + + 2\xi_{1n}\omega_{1n}\dot{q}_{1n} + \omega^2_{1n} q_{1n} - \lambda\alpha_1 \ddot{q}_{2n} = Q_{1n}\sin \omega t.$$

$$\&\ \ddot{q}_{2n} + 2\xi_{2n}\omega_{2n}\dot{q}_{2n} + \omega^2_{1n} q_{2n} - \lambda\alpha_2 \ddot{q}_{2n} = Q_{2n}\sin \omega t.(\omega t + \chi)$$

Where $Q_{jn} = \dfrac{1}{1R_j(m_j + \mu_j M_j)} \int_o^\ell gj(x)\varnothing_{jn}\, dx$ ....(4.32)

The solution of equation (32) are easily obtained

$$q_{1n}^{(t)} = \alpha_{1n}\operatorname{Sin}(wt) + \beta_{1n}\operatorname{Cos}(wt)$$

$$\&\ q_{2n}^{(t)} = \alpha_{2n}\operatorname{Sin}(wt) + \beta_{2n}\operatorname{Cos}(wt)\ \ldots\ldots\ldots\ldots\ldots(4.33)$$

Where $\alpha_{1n}, \alpha_{2n}, \beta_{1n}$ and β_{2n} are solutions of the following equations.

$$\begin{vmatrix} w_{1n}^2 - w^2, & -2\xi_{1n}^w w_{1n}^w & \lambda\alpha_1 w^2 & o \\ 2\xi_{1n}^w w_{1n}^w & w_{1n}^2 - w^2 & o & \lambda\alpha_1 w^2 \\ \lambda\alpha_2 w^2 & 0 & w_{2n}^w - w^2 & 2\xi_{2n}^w 2n^w \\ 0 & \lambda\alpha w^2 & 2\xi_{2n}^w 2n^w & w_{2n}^w - w^2 \end{vmatrix}$$

$$(\alpha_{1n}\ B_{1n}\ \alpha_{2n}\ B_{2n}) = (Q_{1n}\ o\ Q_{2n}\ \operatorname{Cos}\ Q_{2n}) \quad \ldots\ldots\ldots\ldots\ldots(4.34)$$

Substituting equations (4.33) into (4.16) yields

$$\phi_{j(x)} = \left\{ \left[\sum_{n=1}^{\infty} \alpha_{jn} \phi_{jn}(x) \right]^2 + \left[\sum_{n=1}^{\infty} \beta_{jn} \phi_{jn}(x) \right]^2 \right\}^{\frac{1}{2}}$$

and $\Psi_j(x) = \tan^{-1} \left[\dfrac{\sum_{n=1}^{\infty} \alpha_{jn} \phi_{jn}(x)}{\sum_{n=1}^{\infty} \beta_{jn} \phi_{jn}(x)} \right]$ ....(4.35)

CHAPTER-V

VIBRATIONS OF CYLINDERS IN A LIQUID HAVING SQUARE & OTHER GEOMETRICAL SHAPES

5.

Vibrations of Cylinders in a Liquid Having Square and Other Geometric Shapes

In this study, the whole structure composed of many cylinders and a liquid is replaced by a continuum on the assumption of two dimensional motions in order to simplify the analysis. The equations of cylinder motion are derived as partial differential equations to examine the fluid motion and hydrodynamic forces associated with cylinder motion.

The effect of non-linear hydrodynamic force produced by large variation of the gap with between the cylinders, viscosity and compressibility of the fluid are neglected in this analysis.

The natural frequencies are examined by using the Raleigh method (6).

5.1 Basic Equations

5.1.1 Equations of Continuity for Square Cylinders

Two dimensional model of many square cylinders immersed in an incompressible ideal fluid region (**Fig. 5.1**)

The flow velocity at 'x' $'u' = \dfrac{\dot{h}}{h}x + u_0$ ....(5.1)

Where

u_o = flow velocity at x=0

h = gap width

h_o = gap width in equilibrium

$\dot{h}$ = $\frac{dh}{dt}$ differential of gap width with time considering a continuity of flow at the junction.

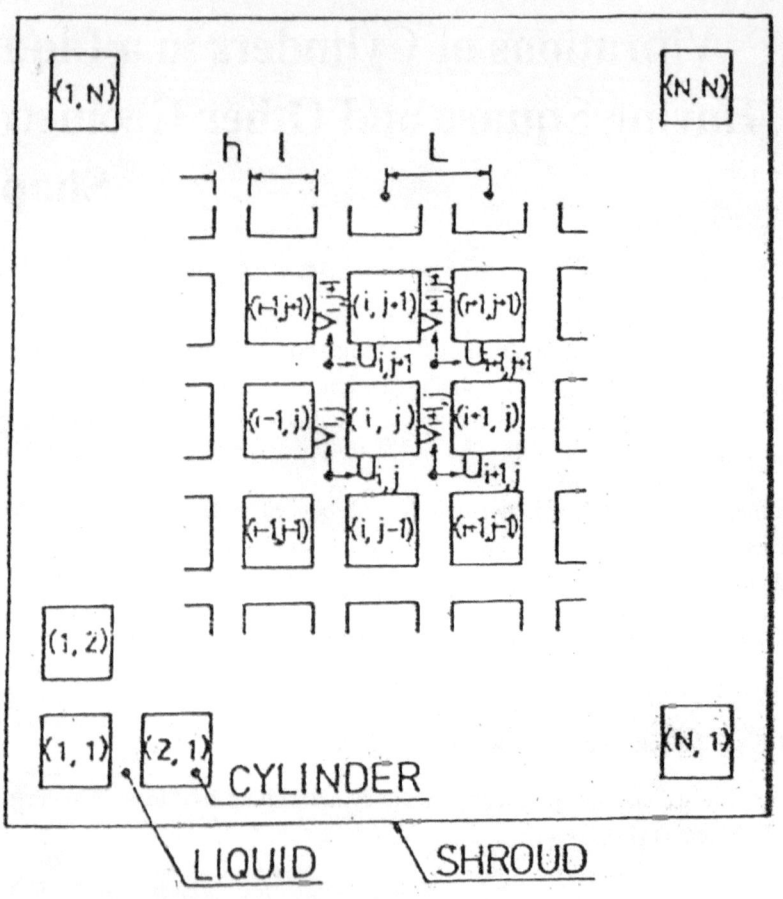

Figure 5.1: Two-Dimensional Model of Square Cylinders in a Liquid

$$U_1 + 1, J + 1 - U_1, JH + v_1 + 1, j + 1 - v_1 + 1, J = -\frac{\ell}{h_o}(h_x, 1j + \dot{h}y\, ij) \quad\quad\quad\quad\quad\quad\quad\quad\quad\quad\quad\quad\quad\quad\quad\quad\quad\quad\quad\quad\quad\quad(5.2)$$

$(i, +j, j, 2 \ldots\ldots N, N^2 =$ total number of cylinder)

Where 1 = length of one side of cylinder

Vibration of Heat Exchanger Tubes

$h_x, ij = h_o + \xi_{i+1,j} - \xi_{ij}$ =gap width between the cylinders $(i+1,j)$ & (i,j)

$h_y, ij = h_o + n_i, j+1 - n_{ij}$ = gap width between the cylinders $(i, j+1)$ and (i, j) as shown in **Fig. 5.1**.

If the number of cylinders N is sufficient large, equation (5.2) is replaced by a partial differential equation. Applying Taylors series expression to $u_{i+1,J+1} - u_{ij+1}, v_{ij+1}, j+1$ $- v_{i+1}, J, \xi_{i+1}, J - \xi_{i1}$ and $\eta_{ij+1} - \eta_{ij}$ and putting

$u_{i+1, j+1}$ = $u(x,y,t)$

$v_{+1, j+1}$ = $v(x,y,t)$

ξ_{ij} = $\xi(x,y,t)$

and η_{i1} = $\eta(x,y,t)$ we obtain the equation of continuity

$$\frac{\partial u}{\partial x} + \frac{\partial v}{\partial y} = -\frac{\ell}{h_o}\left(\frac{\partial^2 \xi}{\partial t \partial x} + \frac{\partial^2 \eta}{\partial t \partial x}\right) \quad \ldots(5.3)$$

Where, u = fluid velocity in x direction

v = fluid velocity in y direction

V = relative velocity of fluid in direction in case of square cylinder.

U = relative velocity of fluid in x direction

I = length of one side of cylinder

c = equivalent damping coefficient of cylinder

F = fluid reaction acting on each cylinder

H = frequency response function

h = gap width between cylinder

ho = h in state of equilibrium

k = equivalent stiffness of cylinder

L = distance between centres of adjacent cylinder

M = mass of cylinder per unit length

N	=	number of cylinders in each row in case of square cylinders.
P	=	pressure
γ_{mm}	=	B_{mn}/A_{mn} = ratio of amplitude of η to that of ξ
t	=	time
X, γ	=	relative displacement in redirection & y direction respectively.
X_f, Y_f	=	displacement of foundation in x&y direction respectively.
X, y	=	coordinates of two dimensional space.
ξ	=	displacement of cylinder in x direction.
η	=	displacement of cylinder inn x direction
ζ	=	equivalent damping ratio of cylinder
∅	=	phase difference between and
$\emptyset_{mn}(x,y)$	=	Sin (mπx /a) Cos (nπy /b)
$\psi_{mn}(x, y)$	=	Cos (mπx /a) Sin (nπy /b)
ρ	=	density of fluid
w	=	circular frequency
w_o	=	$\sqrt{K/M}$ = natural circular frequency of cylinder without coupling effect.
w_{mn}	=	natural frequency of mode (m,n)

5.1.2 Equation of Fluid Motion

Euler's equation of one dimensional fluid motion, pressure P at an arbitrary position x in the represented by suing equation (5.1),

$$p = \frac{\rho}{2ho} - \ddot{h}x^2 - \rho \ddot{u}_o x + p_o \dots\dots\dots\dots(5.4)$$

Applying Taylor's series ad foregoing equations and putting F_x, i, j = F_x (x, y, t) and F_y, i, j = F_y (x, y, t)

# Vibration of Heat Exchanger Tubes

$$Fx = \frac{\rho \ell^2}{6}\left(-\frac{\ell L^2}{h_0}\frac{\partial^4 \xi}{\partial t^2 \partial x^2} - \frac{3\ell L}{h_0}\frac{\partial^3 \eta}{\partial t^2 \partial y} + 3L\frac{\partial^2 v}{\partial t \partial x} + 6\frac{\partial u}{\partial t}\right)$$

$$Fy = \frac{\rho \ell^2}{6}\left(-\frac{\ell^2 L^2}{h_0}\frac{\partial^4 \eta}{\partial t^2 \partial y^2} - \frac{3\ell L}{h_0}\frac{\partial^3 \xi}{\partial t^2 \partial x} + 3L\frac{\partial^2 u}{\partial t \partial y} + 6\frac{\partial v}{\partial t}\right) \quad \text{.........(5.5)}$$

5.1.3 Equation of Cylinder Motion

If every cylinder is assumed to be connected independently with the foundation through linear springs & dampers in x & y, directions then the equation of free vibration of cylinders are reduced to

$$M\ddot{\xi}_{ij} + c\dot{\xi}_{ij} + k\xi_{ij} = Fx, ij$$
$$M\ddot{\eta}_{ij} + c\dot{\eta}_{ij} + k\eta_{ij} = Fy, ij \quad \text{.........(5.6)}$$

$\xi_{ij}, \eta_{ij}, u_{ij}, v_{ij}, F_{xij}$ and F_{yij} are determined for discrete parameter model from above equations

$$\xi_{oj} = \xi_{N+1,j} = u_{ij} = u_{N+1,j} = 0$$
$$\eta_{io} = \eta_{1N+1} = v_{ij} = v_{i,N+1} = 0 \quad \text{.........(5.7)}$$

Replacing equations (5.6) by partial differential equations & further substituting (5.5), we have the equations of free vibration of cylinders as a two dimensional continuum,

$$\left[\frac{\partial^2}{\partial t^2}\left(\frac{\rho l^3 L^3}{6h_0}\frac{\partial^2}{\partial x^2}\right) + C\frac{\partial}{\partial t} + K\right]\xi$$

$$+ \frac{\rho \ell^3 L}{2h_0}\frac{\partial^3 \eta}{\partial t \partial y} - \frac{\rho \ell^2 L}{2}\frac{\partial^2 v}{\partial t \partial x} - \rho \ell^2 L \frac{\partial u}{\partial t} = 0$$

$$\left[\frac{\partial^2}{\partial t^2}\left(\frac{\rho \ell^3 L^2}{6h_0},\frac{\partial^2}{\partial y^2}\right) + C\frac{\partial}{\partial t} + K\right]\eta$$

$$+\frac{\rho\ell^3 L}{2h_o}\frac{\partial^3 \xi}{\partial t \partial y} - \frac{\rho\ell^2 L}{2}\frac{\partial^2 u}{\partial t \partial x} - \rho\ell^2 \frac{\partial v}{\partial t} = 0 \quad \ldots\ldots\ldots\ldots\ldots\ldots\ldots\ldots\ldots\ldots\ldots\ldots\ldots\ldots(5.8)$$

Therefore ξ, η, u and v are determined from equations (5.3), (5.8) under boundary conditions

ξ = u=0 at x = o, a

η = v=0 at y = o, a$\ldots\ldots\ldots\ldots\ldots\ldots\ldots\ldots\ldots\ldots\ldots\ldots$(5.9)

Where a = NL = interior width of square shroud.

5.2 Natural Frequencies

5.2.1 Equations of Energy (Square Cylinders)

To examine approximate values of the natural frequency by using the Rayleigh Method, we introduce energy equation. The potential energy in a square region including single cylinder (ij) is represented by an elastic energy stored in a spring 'k' and kinetic energy in the same square region is produced by the motions of the cylinder and the fluid in the gaps, where the fluid velocities are given by equation (5.1).accordingly, the total energy over the two dimensional continuum region are evaluated as follows.

Potential Energy, $\quad E_s = \dfrac{k}{2L^2}\displaystyle\int_0^a \int_0^a (\xi^2 + \eta^2)\, dx.\, dy \quad \ldots\ldots\ldots\ldots\ldots\ldots\ldots\ldots\ldots\ldots$(5.10)

Kinetic energy of the Cylinders, $\quad E_C = \dfrac{M}{2L^2}\displaystyle\int_0^a \int_0^a \left\{\left(\dfrac{\partial \xi}{\partial t}\right)^2 + \left(\dfrac{\partial \eta}{\partial t}\right)^2\right\} dxdy$

Kinetic energy of the fluid, $E_f =$

$$\frac{\rho\ell}{2h_o L^2}\int_0^a \int_0^a \left\{\frac{\ell^2 L^2}{3}\left(\frac{\partial^2 \eta}{\partial t \partial y}\right)^2 - \ell h_o L \frac{\partial^2 \eta}{\partial t \partial y}, 4 + 2hu + \frac{\ell^2 L^2}{3}\left(\frac{\partial^2 \xi}{\partial t \partial x}\right)^2 - \ell h_o L \frac{\partial^2 \xi}{\partial t \partial x}v + h^2 v^2\right\} dxdy$$

$\ldots\ldots\ldots\ldots\ldots\ldots\ldots\ldots\ldots\ldots\ldots\ldots\ldots\ldots\ldots\ldots\ldots\ldots\ldots\ldots\ldots\ldots\ldots\ldots\ldots\ldots\ldots\ldots\ldots\ldots\ldots\ldots\ldots\ldots\ldots\ldots\ldots\ldots$(5.12)

But $E_s + E_c + E_f$ = constant $\ldots\ldots\ldots\ldots\ldots\ldots\ldots\ldots\ldots\ldots\ldots\ldots\ldots\ldots\ldots\ldots\ldots\ldots\ldots\ldots$(5.13)

Let us presume that ξ, η, u and v are the following functions satisfying the boundary conditions (5.9) and the equation of continuity (5.3)

ξ = $A_{mn}\,\varnothing_{mn}$ (x, y) cos wt $\ldots\ldots\ldots\ldots\ldots\ldots\ldots\ldots\ldots\ldots\ldots\ldots\ldots\ldots\ldots\ldots$(5.14)

η = $B_{mn}\,/_{mn}$ (x, y) cos (wH$\varnothing$)

Vibration of Heat Exchanger Tubes

$$u = \frac{wI}{ho}\frac{n}{m} B_{mn} \varnothing_{mn}(x, y) \sin(wt+\varnothing)$$

$$v = \frac{wI}{ho}\frac{n}{m} A_{mn} \psi_{mn}(x, y) \sin wt \quad \text{................................(5.15)}$$

where A_{mn}, B_{mn} are constants,

$$\varnothing_{mn}(x, y) = \sin\frac{m\pi x}{a} \cos\frac{n\pi y}{a}$$

$$\psi_{mn}(x, y) = \cos\frac{m\pi x}{a} \sin\frac{n\pi y}{a} \quad \text{................................(5.16)}$$

W = circular frequency

$\varnothing$ = phase difference

When $m = 0$, $n \neq 0$ then $\xi = u = 0$

and $\eta = B_{on} \sin\frac{n\pi y}{a} \cos\omega t$

$$v = \frac{\omega I}{ho} B_{on} \sin\frac{n\pi y}{a} \sin \omega t \quad \text{................................(5.17)}$$

When $m \neq 0$, $n = 0$, then $\eta = v = 0$

and $\xi = Am_o \sin\frac{m\pi y}{a} \cos wt$

$$u = \frac{wI}{ho} A_{mo} \sin\frac{m\pi y}{a} \sin wt \quad \text{................................(5.18)}$$

although the assumption of the continuous mode function is valid in the case of the sufficiently large number of cylinders, the discrete mode shape of the finite number of cylinders, is represented by the value of the continuous mode function at the centre of each cylinder, and the maximum mode number is restricted by number of cylinders in a single row.

Substituting foregoing equations into potential energy E_s Kinetic energy E_c, Kinetic energy fluid E_f equations and considering summation of all equal to constant, we get the natural circular frequency,

$$W_{mn} = \cfrac{W_o}{\left\{\cfrac{\left[\left(\dfrac{\ell}{a}\right)^2 \dfrac{\pi^2}{3}(m^2 + n^2\gamma_{mn}\cos 2\emptyset)\right]}{+\left(\dfrac{m}{n}\right)^2 + \left(\dfrac{n}{m}\right)^2 \gamma^2_{mn}\cos 2\emptyset}\right\}^{\frac{1}{2}}}{1 + \cfrac{\rho t^2}{M} x \cfrac{1}{h_o} - x \cfrac{1}{\gamma^2_{mn}\cos 2\emptyset}} \quad \ldots\ldots\ldots(5.19)$$

Where $W_o = \sqrt{k/M}$ = natural circular frequency of the cylinder without coupling effects.

γ_{mn} = B_{mn}/A_{mn} = amplitude ratio .. (m, n = 1, 2, …….)

When m = o, n ≠ o or m ≠ o, n=o or m = n then

$$W_{mn} = W_{mo} = W_{mm} = W_o / \left[1 + \cfrac{\rho \ell^2}{M} \cfrac{\ell}{ho}\left\{\left(\cfrac{L}{a}\right)^2 \cfrac{\pi^2}{3} m^2 + 1\right\}\right]^{\frac{1}{2}}$$

(m = 1, 2, ………..) $\quad\ldots\ldots\ldots\ldots\ldots\ldots\ldots\ldots$(5.20)

It is seen from the forgoing equations, that the natural frequency is low, if the mass ratio = $\dfrac{\rho l^2}{M}$ increases and the gap ratio ho/ℓ decreases. In otherworld, the virtual mass coefficient is proportional to $\dfrac{\rho l^2}{M}$ and ho/ℓ. As a matter of course, the natural frequency equal to that of single cylinder if the fluid density tends to zero or the gap width tends to infinity. In the case of above equation (5.20), the natural frequency decreases as the ratio of the mode number 'm' to the number of cylinder n= a/L increases.

5.2.2 Energy Equations for Hexagonal Cylinder

Considering the condition of constant of total energy represented by these mode functions, we get the natural circular frequency, when m ≠ o, and n = o

$$W_{mo} = W_{o/}\left[\cfrac{\sqrt{3\pi^2}}{36}\left(\cfrac{\sqrt{3\rho l^2}}{2M}\right)\cfrac{\ell}{ho}\left(\cfrac{L}{a}\right)^2 m^2 + \sqrt{3}\left(\cfrac{3\sqrt{3\rho l^2}}{2M}\right)\cfrac{\ell}{ho} + 1\right]^{\frac{1}{2}} \ldots\ldots\ldots(5.21)$$

When m = o, and n ≠ o

$$W_{mo} = W_{ol}\left[\frac{\sqrt{3}\pi^2}{12}\left(\frac{3\sqrt{3}\rho l^2}{2M}\right)\frac{\ell}{ho}\left(\frac{L}{b}\right)^2 m^2 + \sqrt{3}\left(\frac{3\sqrt{3}\rho \ell^2}{2M}\right)\frac{\ell}{ho}+1\right]^{\frac{1}{2}}$$

(m = 1, 2, ……..) ……………………………………………..(5.22)

As described later, ω_{mo} and ω_{om} are the uncoupled circular frequencies of the vibrations in the x and y directions respectively. Since ω_{mo} and ω_{om} are from above two equations (5.21) and (5.22), the natural frequency of the vibration in x, directions is higher than that in the y-direction. This fact may come from difference between the directions of flow channel to the x & y axes.

In the case of hexagonal cylinders, the virtual mass coefficient is proportional to $3\sqrt{3}\rho\ell^2/2M$ and ℓ/ho. The natural frequency decreases as the number of cylinder a/L or b/L increases.

When a/L, b/L $\to \infty$ m=0, n $\neq$0 or m $\neq$ 0, n=0 above equations, become.

$$W_{mo} = W_{om} = \sqrt{\frac{K}{M+(9\rho\ell^3/2ho)}}$$

$$= w/\sqrt{1+(\sqrt{3}(3\sqrt{3}\rho\ell^2/2M)(\ell/ho)}\ldots\ldots\ldots\ldots\ldots\ldots\ldots\ldots\ldots\ldots\ldots(5.23)$$

Where the factor $3\sqrt{3}/2M$ represents the ratio of the fluid mass replaced by the cylinders to the cylinder mass.

5.3 Frequency Response

For the response analysis of the cylinders under stationary excitation, the frequency response should be known. We hereby consider square cylinder only.
The frequency response function of X_{mo} to the input acceleration $\ddot{X}_f$ is given as follows.

$$Hm(jw) = -\frac{4}{m\pi}\left(1-\frac{\rho l^2}{M}\right)l\left\{-\left[1-\frac{\rho l^2}{M}\frac{\ell}{ho}\left\{\frac{\pi}{6}\left(\frac{L}{a}\right)\right\}^2 m^2-1\right]w^2+2\xi w_o.jw+w_o^2\right\}$$

……………………………………………..(5.24)

Where, m = odd

$\xi = c/2\sqrt{KM}$ = damping ratio of cylinder without coupling.

The frequency response function of γ_{on} (n = odd) to the input acceleration $\ddot{Y}_f$ is also given by this equation. From this, it is seen that the relative displacement is not produced by the uniformly distributed input acceleration, if the fluid mass replaced by the cylinder is equal to the cylinder mass, $M = \rho \ell^2$.

CHAPTER-VI

DYNAMIC RESPONSE OF THREE PARALLEL CIRCULAR CYLINDERS IN A LIQUID

6.

Dynamic Response of Three Parallel Circular Cylinders in a Liquid

The dynamic responses of two parallel circular cylindrical rods were seen in Chapter IV. The work was carried out by S.S. Chen [5] now author wants to develop the mathematic approach for dynamic response of three parallel circular cylinders in a liquid and also try to see the hydrodynamic effect and develop the equation of motion with added mass coefficient concept. It would be seen the gap-radius ratio against added mass coefficient finally the frequency response with exact solution method considering free vibration of cylinders in fluid flow when damping and forcing factor neglected. The approach for approximate solution and also for forced vibration can also be developed in similar method. The results of this study is to find out or try to open an approach achieve maximum efficiency of Heat Exchangers under fluid flow excited vibrations and what should be the best possible gap-between cylinders, if they are of same radii.

6.1 Dynamic Response of Three Parallel Circular Cylindrical Tubes

Consider three parallel circular cylindrical tube '1', '2', & '3' as shown in **Fig. 6.1** are immersed in a liquid where, R_{1i}, R_{2i}, R_{3i} = Internal Radius of first 1st, 2nd and 3rd tubes respectively and R_{1o}, R_{2o}, R_{3o}, outer radius of 1st, 2nd & 3rd tubes respectively.

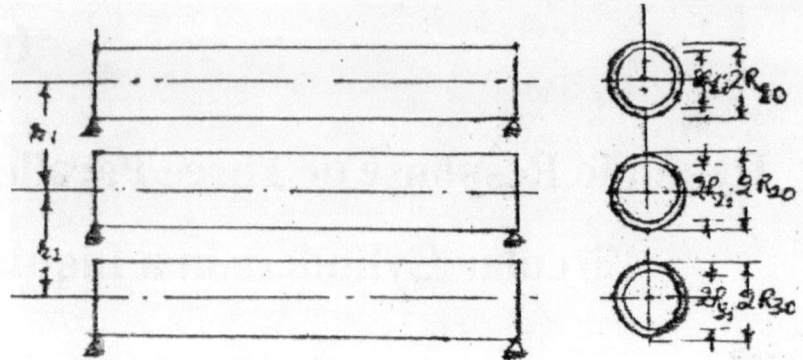

Figure 6.1: Three Parallel Circular Cylindrical Tubes in a Liquid

Tubes motions consist of an in plane displacement along y axis and an out of plane displacement along the 'Z' axis. The equation of motion for either in plane or out of plane motion can be written.

$$E_1 I_1 \frac{\partial^4 u_1}{\partial x^4} + C_1 \frac{\partial u_1}{\partial x} + M_1 \frac{\partial^2 u_1}{\partial t^2} = (F_{10} - F_{1i}) + f_1$$

$$E_2 I_2 \frac{\partial^4 u_2}{\partial x^4} + C_2 \frac{\partial u_2}{\partial t} + M_2 \frac{\partial^2 u_2}{\partial t^2} = (F_{20} - F_{2i}) + f_2$$

$$E_3 I_3 \frac{\partial^4 u_2}{\partial x^4} + C_3 \frac{\partial u_3}{\partial t} + M_3 \frac{\partial^2 u_3}{\partial t^2} = (F_{30} - F_{3i}) + f_3$$

$$\dots\dots\dots\dots\dots\dots(6.1)$$

In general, if there are 'n' number of tubes in the system then the equation motion can be represented as under,

$$E_j I_j \frac{\partial^4 u_j}{\partial x^4} + C_j \frac{\partial u_j}{\partial t} + M_j \frac{\partial^2 u_j}{\partial t^2} = (F_{j0} - F_{ji}) + f_j \dots\dots\dots\dots(6.2)$$

There j = 1, 2, 3 ....n for corresponding cylindrical tubes.

The hydrodynamic forces associated with two circular vibrating rods were given by Mazur ref (5), using two dimensional theory,

Vibration of Heat Exchanger Tubes

$$F_1 = -M_1\ \mu_{11}\ \frac{\partial^4 u_1}{\partial t^2} + -M_1\ \mu_{12}\ \left(\frac{R2}{r_1}\right)^2 \frac{\partial^2 u_2}{\partial t^2}$$

$$F_2 = -M_2\ \mu_{22}\ \frac{\partial^2 u_2}{\partial t^2} + M_2\ \mu_{21}\ \left(\frac{R1}{r_1}\right)^2 \frac{\partial^2 u_1}{\partial t^2} \quad \ldots\ldots\ldots\ldots\ldots\ldots(6.3)$$

This equation represents that the hydrodynamic force is due to acceleration corresponding tube '1` + associated force due to acceleration with coupling effect corresponding to rod/tube '2'. Secondly the hydrodynamic force is inversely proportion to the gap between two cylinders. Thus, in case gap increases the 2nd function of equation goes on reducing (i.e. coupling effect reduces)

6.1.1 Nomenclatures

Following nomenclatures are going to be used for deriving the equation of motions.

F_{jo} = Hydrodynamic force on the tube in j th tube

F_{ji} = Hydrodynamic force on account of the fluid following inside the tube.

M_j = Displacement of mass of fluid by tube per unit length.

M_j = mass of tube per unit length.

I_j = moment of inertia of tube.

C_j = Coefficient of viscous damping.

f_j = excitation force.

r_j = distance between centres of two adjacent cylindrical tubes.

R_{jo} = tube outer radius

R_{ji} = tube inner radius

t = tune

u_j = tube displacement

μ_{jn} = added mass coefficient of tube 'jth' in respect of tube 'n'.

ρ = fluid density.

W = mass of fluid flowing inside the tube per unit length.

E = modules of Elasticity of tube material.

6.1.2 Assumptions

Now we are going to consider following assumptions:

(i) the hydrodynamic effect will be only due to neighbouring cylinders.

(ii) The distance between centre of the two consequent cylinders are same as 'r' = $r_1 = r_2 = ..rj$

(iii) R< 8R as beyond this there is no effect of coupling ref. (4).

Since hydrodynamic forces for three cylinders (in plane can be written as

$$F_1 = -M_1 \, \mu^o{}_{12} \frac{\partial^2 u_1}{\partial t^2} + M_1 \, \mu_{12} \left(\frac{R_2}{r}\right)^2 \frac{\partial^2 u_2}{\partial t^2}$$

$$F_2 = -M_2 \, \mu^o{}_{21} \frac{\partial^2 u_2}{\partial t^2} + M_2 \, \mu_{21} \left(\frac{R_1}{r}\right)^2 \frac{\partial^2 u_1}{\partial t^2}$$

$$- M_2 \, \mu^o{}_{23} \frac{\partial^2 u_2}{\partial t^2} + M_2 \, \mu_{23} \left(\frac{R_3}{r}\right)^2 . \frac{\partial^2 u_3}{\partial t^2}$$

$$F_3 = -M_3 \, \mu^o{}_{32} \frac{\partial^2 u_3}{\partial t^2} + M_3 \, \mu_{32} \left(\frac{R_2}{r}\right)^2 \frac{\partial^2 u_2}{\partial t^2} \quad \ldots\ldots(6.4)$$

Where (in general considering pth & qth tubes

$$\mu^o pq = 1 + \frac{r^2 - 2r^2(Rp^2 + Rq^2) + (Rq^2 + Rp^2)}{r^2 Rp^2}$$

$$\sum_{k=1}^{\infty} k \frac{e^{-k(h+hp)}}{Sinh(kh)}$$

$$\mu^o pq = 1 + \frac{r^4 - 2r^2(Rp^2 + Rq^2) + (Rq^2 + Rp^2)^2}{Rp^2 Rq^2}$$

$$\sum_{k=1}^{\infty} h\, Coth(kh)\, e^{(-2kh)} \quad \ldots\ldots\ldots\ldots\ldots(6.5)$$

iff,

$$h = \ln\left\{\frac{r^2 - Rp^2 - Rq^2}{2Rp\, Rq} + \left[\left(\frac{r^2 - Rp^2 - Rq^2}{2Rp\, Rq}\right)^2 - 1\right]^{\frac{1}{2}}\right\}$$

$$hp = 2\ln\left\{\frac{r^2 + Rp^2 - Rq^2}{2r\, Rq} + \left[\left(\frac{r^2 + Rp^2 - Rq^2}{2r\, Rq}\right)^2 - 1\right]^{\frac{1}{2}}\right\}$$

$$\text{and } h_q = 2\ln\left\{\frac{r^2 - Rp^2 - Rq^2}{2r\, Rq} + \left[\left(\frac{r^2 - Rp^2 + Rq^2}{2r\, Rq}\right)^2 - 1\right]^{\frac{1}{2}}\right\} \ldots\ldots\ldots(6.6)$$

Introducing, $Rp = Rq = R$ (say) when all tubes are same diameter and uniform mass per unit length (isotropic),

$$\mu^0_{pq} = \mu^0_{pq} = 1 + \frac{r^4 - 4r^2 R^2}{r^2 R^2} \sum_{k=1}^{\infty} k \frac{e^{[-k(h+hp)]}}{Sinh(kh)}$$

$$\mu_{pq} = \qquad 1 + \frac{r^4 - 4r^2 R^2}{2R^4} \sum_{k=1}^{\infty} h\, Coth(kh)\, e(-2kh) \quad \ldots\ldots\ldots\ldots(6.7)$$

Where $h = \ln\left\{\dfrac{r^2 - 2R^2}{2R^2} + \left[\left(\dfrac{r^2 - 2R^2}{2R^2}\right) - 1\right]^{\frac{1}{2}}\right\}$

$$hp = hq = 2\ln\left\{\dfrac{r^2}{2R^2} + \left[\left(\dfrac{r}{2R}\right) - 1\right]^{\frac{1}{2}}\right\} \quad \ldots\ldots(6.8)$$

Now, again introducing,

Hp = hq = h*

$\mu^o{}_{pq} = \mu^o{}_{qp} = \mu^*$

and $\mu_{pq} = \mu_{qp} = \mu,$ ....(6.9)

We can re-write the hydrodynamic forces, as,

$$F1 = M_1\mu^* \dfrac{\partial^2 u_1}{\partial t^2} + M_1 \mu \left(\dfrac{R}{r}\right)^2 \dfrac{\partial^2 u_2}{\partial t^2}$$

$$F2 = M_2\mu^* \dfrac{\partial^2 u_2}{\partial t^2} + M_2 \mu \left(\dfrac{R}{r}\right)^2 \dfrac{\partial^2 u1}{\partial t^2} + \dfrac{\partial^2 u3}{\partial t^2}$$

$$\& F3 = M_3\mu^* \dfrac{\partial^2 u_3}{\partial t^2} + M_3 \mu \left(\dfrac{R}{r}\right)^2 \dfrac{\partial^2 u2}{\partial t^2} \quad \ldots\ldots(6.10)$$

Where,

$$\mu^* = 1 + \dfrac{r^4 - 4r^2 R^2}{2R^4} \sum_{k=1}^{\infty} k \dfrac{e^{-k(h + h^*)}}{\text{Sinh}(kh)}$$

$$\mu = 1 + \dfrac{r^4 - 4r^2 R^2}{2R^4} \sum_{k=1}^{\infty} h\coth(kh) e^{(-2kh)} \quad \ldots\ldots(6.11)$$

and, $h = In\left\{\dfrac{r^2 - 2R^2}{2R^2} + \left[\rho\left(\dfrac{r^2 - 2R^2}{2R^2}\right)^2 - 1\right]^{\frac{1}{2}}\right\}$

$h^* = 2In\left\{\dfrac{r}{2R} + \left[\left(\dfrac{r}{2R}\right)^2 - 1\right]^{\frac{1}{2}}\right\}$ ....(6.12)

From the equations (6.10), it reveals that hydrodynamic forces F_1 F_3 are identical. Thus, it would be concluded that hydrodynamic forces for extreme outer tubes facing shell, if they are fitted in Heat exchanger tube bundle, are similar, but it is different than the forces for intermediate tubes. Thus, dynamic responses of intermediate tubes are very important to analyse the heat exchanger vibration.

Substituting the values of F_1, F_2 and F_3 from equation (6.10) in (6.1) and $(F_1 - F_{i1}) = F_1$, $(F_{02}-F_{i2}) = F_3$ and $(F_{03}-F_{i2}) = F_3$, we get

$$E_1 I_1 \dfrac{\partial^4 u1}{\partial x^4} + C_1 \dfrac{\partial u1}{\partial t} + (m_1 + \mu * M_1)\dfrac{\partial u_1}{\partial t^2} - \lambda\mu M_1 \left(\dfrac{R}{r}\right)^2 \dfrac{\partial^2 u2}{\partial x^2} = f1$$

$$E_2 I_2 \dfrac{\partial^4 u2}{\partial x^4} + C_2 \dfrac{\partial u2}{\partial t} + (m_2 + 2\mu * M_2)\dfrac{\partial^2 u2}{\partial t^2} - \lambda\mu M_2 \left(\dfrac{R}{r}\right)^2$$

$$\left[\dfrac{\partial^2 u1}{\partial t^2} + \dfrac{\partial^2 u3}{\partial t^2}\right] = f2$$

$$E_3 I_3 \dfrac{\partial^4 u3}{\partial x^4} + C_3 \dfrac{\partial u2}{\partial t} + (m_3 + 2\mu * M3)\dfrac{\partial^2 u3}{\partial t^2} - \lambda\mu M_3 \left(\dfrac{R}{r}\right)^2 \dfrac{\partial^2 u2}{\partial t^2} = f3 \ldots(6.13)$$

Where $\lambda=1$ for in plane motion and $\lambda= -1$ for out of plane motion. Cylinder motions in the two planes are uncoupled.

6.2 Free Vibration

To find out the vibrations frequency first, we shall adopt the exact solution method. Neglect the damping & forcing factor in equation (6.13) and let

$$U_j = R_j \ddot{u}_{je}^{iwt} \quad \text{...............(6.14)}$$

where I $= \sqrt{-1}$

w $=$ vibration frequency

j $=$ 1, 2, 3 for corresponding tubes

thus, $u_1 = \overline{R_1 u_1} e^{iwt} = \overline{R_1 u_1}\{\cos(wt) + i\sin wt)\}$ (Fourier Series)

Using four series and differentiating this equation

Then $\quad \dfrac{\partial^4 u_1}{\partial x^4} + R_1 e^{iwt} \dfrac{\partial \overline{u}1}{\partial x^4}$

Similarly $\quad \dfrac{\partial^2 u_1}{\partial t^2} - R_1 \overline{u}1\, w^2 e^{iwt}$

$$\dfrac{\partial^2 u_3}{\partial t^2} - R_3 \overline{u}_3 W^2 e^{iwt} \quad \text{...............(6.15)}$$

Substituting these values in equation (6.15), we get

$$E_1\, I_1 R_1 e^{iwt} \dfrac{\partial^4 \overline{u}1}{\partial x^4} - (m_1 + \mu * M_1) R_1 \overline{\mu}_1\, w^2 e^{iwt}$$

$$+ \lambda M_1 \mu \left(\dfrac{R}{r}\right) \overline{u}_2\, R_2\, w^2 e^{iwt} = 0$$

$$E_2\, I_2 R_2 e^{iwt} \dfrac{\partial^4 \overline{u}2}{\partial x^4} - (m_2 + 2\mu * M_2) R2 \overline{\mu}_2\, w^2 e^{iwt}$$

$$+ \lambda M_2 \mu \left(\dfrac{R}{r}\right)^2 R_1 \overline{u}_1\, w^2 e^{wt} + R_3 \overline{u}_3\, w^2 e^{iwt} = 0$$

$$E_3 I_3 R_3 e^{iwt} \frac{\partial^4 \bar{u}3}{\partial x^4} - (m_3 + \mu * M_3) R_3 \bar{u}_3 w^2 e^{iwt}$$

$$+ \lambda M_3 \mu \left(\frac{R}{r}\right)^2 \bar{u}_2 R_2 w^2 e^{wt} = 0$$

Re-writing & substituting $R_1 = R_2 = R_3 = R$, we get

$$\frac{\partial^4 \bar{u}_1}{\partial x^4} - \frac{m_1 + \mu * M_1}{E_1 I_1} w^2 \bar{u}_1 + \frac{\lambda M_1 \mu \left(\frac{R}{r}\right)^2 w^2}{E_1 I_1} \bar{u}_2 = 0$$

$$\frac{\partial^4 \bar{u}_2}{\partial x^4} - \frac{m_2 + \mu * M_2}{E_2 I_2} w^2 \bar{u}_2 + \frac{\lambda M_2 \mu \left(\frac{R}{r}\right)^2 w^2}{E_2 I_2} (\bar{u}_1 + \bar{u}_3) = 0$$

$$\frac{\partial^4 \bar{u}_3}{\partial x^4} - \frac{m_3 + \mu * M_3}{E_3 I_3} w^2 \bar{u}_3 + \frac{\lambda M_3 \mu \left(\frac{R}{r}\right)^2 w^2}{E_3 I_3} \bar{u}_2 = 0 \quad \ldots\ldots(6.16)$$

Assuming materials of tubes are isotropic & homogeneous,

as $R_1 = \quad R_2 R_3 = R = m$, therefore

$m_1 = m_2 = m_3 = M$ (say)

$M_1 = M_2 = M_3 = M$ (say)

& Hence $E_1 I_1 = E_2 I_2 = E_3 I_3 = E_1$ (say)$\ldots\ldots\ldots\ldots\ldots\ldots\ldots\ldots\ldots\ldots$(6.17)

Substituting these values in equations (6.16), if reduces as

$$\frac{\partial^4 u_1}{\partial x^4} - \frac{m + \mu * M}{EI} w^2 \bar{u}_1 + \frac{\lambda M \mu \left(\frac{R}{r}\right)^2}{EI} w^2 \bar{u}_2 = 0$$

$$\frac{\partial^4 u_2}{\partial x^4} - \frac{m+\mu^*M}{EI} w^2 \bar{u}_2 + \lambda M\, \mu\left(\frac{R}{r}\right)^2 w^2 (u_1 + u_3) = 0$$

$$\frac{\partial^4 u_3}{\partial x^4} - \frac{m+\mu^*M}{EI} w^2 \bar{u}_3 + \frac{\lambda M\, \mu\left(\frac{R}{r}\right)^2}{EI} w^2 \bar{u}_3 = 0 \quad\ldots\ldots\ldots(6.18)$$

Again introducing the following terms

$$\propto = -\frac{m}{EI}$$

$$\beta = -\frac{\mu^*M}{E\bar{h}}$$

$$\lambda = \frac{\lambda M\, \mu\left(\frac{R}{r}\right)^2}{EI} \quad\ldots\ldots\ldots(6.19)$$

Therefore, equation (6.18) can be re-written as

$$\frac{\partial^4 u_1}{\partial x^4} + (\propto + \beta) w^2 \bar{u} + v w^2 \bar{u}_2 = 0$$

$$\frac{\partial^4 u_2}{\partial x^4} + (\propto + 2\beta) w^2 \bar{u}_2 + v w^2 (u_1 + u_3) = 0$$

$$\frac{\partial^4 u_2}{\partial x^4} + (\propto + \beta) w^2 \bar{u}_2 + v w^2 u_2 = 0 \quad\ldots\ldots\ldots(6.20)$$

The solution of equation (6.20), assuming

$$\bar{u}_1 = \sum_{n=1}^{\infty} a_{ne}(P_{\bar{n}}^{\,x})$$

$$\bar{u}_2 = \sum_{n=1}^{\infty} a_n \eta_n e(P_{\dot{n}}^x)$$

$$\bar{u}_3 = \sum_{n=1}^{\infty} a_n \xi_n e\, P_{\dot{n}}(x) \quad \dotfill (6.21)$$

Where $a_n \eta_n \xi_n$ are arbitrary constants

Now differentiating equation (6.21), we get

$$\frac{\partial^4 u_1}{\partial x^4} = \sum_{n=1}^{\infty} a_n P_n^4 e\, P_{\dot{n}}^x$$

$$\frac{\partial^4 u_2}{\partial x^4} = \sum_{n=1}^{\infty} a_n \eta_n P_n^4 e\, P_{\dot{n}}^x$$

$$\frac{\partial^4 u_3}{\partial x^4} = \sum_{n=1}^{\infty} a_n \xi_n\, P_n^4 e\, P_{\dot{n}}^x \quad \dotfill (6.22)$$

Where $\eta_n = \dfrac{\bar{u}_2}{\bar{u}_1}, \xi_n = \dfrac{\bar{u}_3}{\bar{u}_1}$ \quad \dotfill (6.23)

Substituting (6.22) in (6.21), we get

$$P_n^4 \bar{u}_1 + (\infty + \beta)\, \bar{u}_1 w^2 + \gamma \bar{u}_2 w^2 = 0$$

or $\dfrac{\bar{u}_2}{\bar{u}_1} = \eta_4 = \dfrac{(\infty + \beta) w^2 + P_n^4}{\gamma w^2} = 0$ \quad \dotfill (6.24)

Further,

$$P_n^4 \bar{u}_2 + (\infty + 2\beta)\, \bar{u}_2 w^2 + \gamma (\bar{u}_1 + \bar{u}_3) w^2 = 0$$

or $\{P_n^4+w^2(\propto+2\beta)\}\left\{\dfrac{P_n^4+w^2(\propto+\beta)}{Vw^2}\right\}\bar{u}_1+Vw^2\bar{u}_1+\bar{u}_1+\gamma w^2\bar{u}_3=0$

or $[\{P_n^4+(\propto+2\beta)w^2\}\{P_n^4+(\propto+\beta)w^2\}-\gamma^2w^2]\bar{u}_1=\gamma^2w^4\bar{u}_3=0$

or $\dfrac{\bar{u}_3}{\bar{u}_1}=\xi_n=\dfrac{v^2w4}{[\{P_n^4+(\propto+2\beta)w^2\}\{P_n^4+(\propto+\beta)w^2\}-\gamma^2w^2]}$ ....(6.25)

Again,

$\dfrac{\partial^4 u_3}{\partial x^4}=w^2(\propto+\beta)\bar{u}_3+w^2\gamma u_2=0$

$\displaystyle\sum_{n=1}^{\infty} a_n\xi_n\, P_n^4 e\, P_n^x+w^2(\propto+\beta)\bar{u}_3+w^2\gamma\bar{u}_2=0$

or $P_n^4\,\bar{u}_3+w^2(\propto+\beta)\bar{u}_3+w^2\gamma\bar{u}_2=0$

Substituting the values of u3, u2 from equations (6.24) and (6.25) we get

$\{P_n^4+w^2(\propto+\beta)\}x\dfrac{v^2w4}{[\{P_n^4+w^2(\propto+2\beta)w^2\}\{P_n^4+w^2(\propto+\beta)\}]-\gamma^2w^4}\bar{u}_1$

$=\dfrac{\{P_n^4+w^2(\propto+\beta)\}}{\gamma w^2}\bar{u}_1$

or $\{P_n^4+w^2(\propto+2\beta)\}\{P_n^4+w^2(\propto+\beta)]-\gamma^2w^4\}-v^3w^6=0$

or $\{P_n^4+w^2(\propto+2\beta)\}\{P_n^4+w^2(\propto+\beta)\}-P_n^4\gamma^2w^2-\propto\gamma^2w^6$

$2\beta\gamma^2w^6-\gamma^3w^6=0$

or $P_n^8+P_n^4(\propto+2\beta)w^2+P_n^4(\propto+\beta)w^2+(\propto+\beta)+(\propto+2\beta)w^4$
$-P_n^4\gamma^2w^2-(\propto+2\beta+\gamma)\gamma^2w^6=0$

or
$$P_n^8 + P_n^4(2\alpha + 3\beta)w^2 - \{P_n^4\gamma^2 + (\alpha + \beta)(\alpha + 2\beta)\}w^4$$
$$- (\alpha + 2\beta + \gamma)\gamma^2 w^6 = 0 \qquad \ldots\ldots\ldots\ldots\ldots(6.26)$$

This is the vibration frequency equation for the three parallel cylinders immersed in a liquid. The approach for evaluating the frequency for forced vibration can also be adopted as seen in **Chapter-IV**.

CHAPTER-VII

RESULTS AND DISCUSSION

7. Results and Discussion

7.1 Two Parallel Circular Cylindrical Rods Immersed in a Liquid

(a) **From the study of chapter IV**, it is evident that the boundary conditions of two rods are different; the axial mode shapes of the individual rods during coupled rod motion will be different from those of the individual rods in vacuo. However, if the rods have the same type of boundary and are of the same length, the axial mode shapes of the coupled modes will be the same as the individual rods in vacuo.

(b) **From equations**, (4.21) and (4.22),
Where

$$\begin{vmatrix} A & B \\ C & D \end{vmatrix} = 0 \quad \quad \quad \text{.................(4.25)}$$

Where

$A_{mn} = (w^2 - w^2 n)\delta_{mn}$
$B_{mn} = -\lambda \propto_1 w^2 a_n^m$
$C_{mn} = -\lambda \propto_2 w^2 a_n^m$
$D_{mn} = (w^2 - w^2{}_{2n})\delta_{mn}$

and frequency equation of (4.25) becomes,

$$(1 - \propto_1 \propto_2) w^2 - (w^2{}_{1n} + w^2{}_{2n}) w^2 w^2{}_{1n} w^2{}_{2n} = 0$$

We see that the frequencies are independent of 'x', therefore, the frequencies of in plane motion and cut of plane are the same.

In Ω_{In}, $\bar{q}_{2n}/\bar{q}_{1n}$ is negative for in plane motion ($\lambda=1$) and is positive for out of plane motion ($\lambda=-1$) : while Ω_{2n}, $\bar{q}_{2n}/\bar{q}_{1n}$ is positive for in plane motion and negative for out of plane motion. That is, there exist out of phase modes in which rods more in opposite directions and in phase modes when rods moves in the same directions. Furthermore, the frequency of the out of mode of in plane motion is same as that of in phase mode of out of plane motion, while the frequency of the in phase mode of in plane motions is the same as that of the out of phase mode of plane motion (**Fig. 7.1**).

(c) From equations (4.28) $\Omega_{In} < W_{In}, W_{2n}$

and $\Omega_{2n} > W_{In}, W_{2n}$

It is observed that when two rods are not identical the frequency obtained by assuming that the rod with higher natural frequency is rigid, is the upper bound of frequency of out of phase mode of in plane motion and –in-phase mode of out of plane motion, while the frequency obtained

By assuming that the rods with lower natural frequency are rigid, is the lower bound of the frequency of in-phase mode of in-plane motion and out of phase motion.

(d) The coupling between two rods depends on the radios ratio = R_2/R_1, gap radius ratio = G/R_1 and mass ratio = β_j is small and G/R_j is large, the two rods will respond in-dependently.

(e) When two rods have the same type of boundary condition the exact solution for frequencies is obtained in closed form s given in equation (4.26)

$$\Omega_{In=} \frac{\left\{(w_{1n}^2 + w_{2n}^2) - \left[(w_{1n}^2 - w_{2n}^2) + 4\alpha_1\alpha_2\, w_{1n}^2 - w_{2n}^2\right]^{\frac{1}{2}}\right\}^{\frac{1}{2}}}{[2(1-\alpha_1\alpha_2)]^{\frac{1}{2}}}$$

and

$$\Omega_{In=} \frac{\left\{(w_{1n}^2 + w_{2n}^2) + \left[(w_{1n}^2 - w_{2n}^2) + 4\alpha_1\alpha_2\, w_{1n}^2 - w_{2n}^2\right]^{\frac{1}{2}}\right\}^{\frac{1}{2}}}{[2(1-\alpha_1\alpha_2)]^{\frac{1}{2}}}$$

Vibration of Heat Exchanger Tubes

for rods with different end conditions, the frequencies may be computd from the exact frequency equation (4.12)

$F(w, E_j\ I_j\ m_j, R_j, I\ \rho, G) = 0$

and from approximate solution, equation (4.21) & (4.22)

$$\left| \frac{A}{C} + \frac{B}{D} \right|$$

Where

$$A_{mn} = (w^2 - w^2 1n)_{mn}$$
$$B_{mn} = -\lambda\ \alpha_1\ w^2 a_{nm}$$
$$C_{mn} = -\lambda\ \alpha_2\ w^2 a_{nm}$$
$$D_{mn} = (w^2 - w^2{}_{2n})\delta_{mn}$$

From **Fig.** (7.2), it is clear that when two identical rods or steel tubes vibrating in water, the frequencies of coupled modes is calculated from those of the individual rods considerably when G/R_1 is small. As G/R_1 increases, the interactions between the two rods verses fluid become small and for large G/R_1, they will respond independently of each other.

7.2 Square Cylinders Immersed in a Liquid

From the study of **Chapter V**, it reveals that
(a) the natural frequency as per equation (5.20)

$$w_{om} = w_{mo} = w_{mm} = w_o \Bigg/ \left[1 + \frac{\rho \ell^2}{M}\frac{\ell}{ho}\left\{ \left(\frac{\ell}{a}\right)\frac{2\pi^2}{3}m^2 + 1 \right\} \right]$$

For the mode number m= o, n=1
or   m=1, n=0
are as per the graphical representation **Fig.** (7.3) in which w_{mn}/w_o is plotted against a dimensionless parameter

$(\zeta \ell^2 / m)\left(\dfrac{\ell}{ho}\right)$ by taking the cylinder number in each row N= a/L as a parameter when N→∞, the curves asymptotically converge to the expression

$$W_{om} = W_{mo} = W_{mm} = \sqrt{\dfrac{K}{M + (\lambda \ell / ho)}} \quad \ldots\ldots\ldots\ldots\ldots\ldots\ldots\ldots\ldots(7.1)$$

and the natural frequency is about 30% of that of a cylinder without coupling effects for $(\rho \ell^2 / M)(\ell / ho = 10)$

When gap ratio = 0.1 and mass ratio = 1

For $(\rho \ell^2 / M)(\ell / ho) = 1.0$, natural frequencies are 70%

and $(\rho \ell^2 / M)(\ell / ho) = 0.1$, Natural frequencies are 95%

(b) In the discrete model, the equation of Eigen Values is derived from equations Where, i, j = 1, 2 ....N

$$U_{i+1,j+1} - U_{ij} + 1 + V_{i+1} - V_{i+1}j = -\dfrac{\ell}{ho} \cdot (h_{x,}ij + h_{y,}ij)$$

$$= \dfrac{\ell}{2ho}\{(\ddot{h}_{xij} - \ddot{h}x_i - 1j) - (\ddot{h}_{yij} - \ddot{h}_{yij-1})\}$$
$$(\dot{u}_{ij} - \dot{u}_{ij} + 1) - (v_{ij} - \ddot{h}_{i+1,j})$$

$$F_{x,ij} = \dfrac{\rho \ell^3}{6ho}(\ddot{h}_{xi-1,j} - \ddot{h}_{x,ij}) - \dfrac{\rho \ell^2}{2}(\dot{v}_{ij} - \dot{v}_{i+j-j}) + \ell(p_{ij} - pi+1,j)$$

$$F_{y,ij} = \dfrac{\rho \ell^3}{6ho}(\ddot{h}_{y,i,j-1} - \ddot{h}_{y,i,j}) - \dfrac{\rho \ell^2}{2}(\dot{u}_j - \dot{u}_{i,j+1}) + \ell(P_{ij} - P_{ij+1})$$

and $\quad M\ddot{\xi}_{ij} + c\dot{\xi}_{ij} + k\xi_{ij} = F_{x,3j}$

$\quad M\ddot{\eta}_{ij} + c\dot{\eta}_{ij} + k\eta_{ij} = F_{x,ij}$

& by neglecting the damping coefficient. The lowest natural frequency is plotted against $(\rho \ell^2 / M)\left(\dfrac{\ell}{ho}\right)$ for N = 1, 2, 3 and 4 in (**Fig. 7.4**).

It is seen from this numerical results that the natural frequency for the small number of cylinders is higher than that of many cylinders. And it may be expected that the natural frequency approaches that of equation (7.1), when the cylinder number becomes sufficiently large. Therefore, the analytical results obtained with the aid of the continuum model may be valid with good accuracy in the case of many cylinders.

(c) When the damping ratio is zero, the resonance circular frequency is determined by equation (5.24), as given below,

$$w = wo / \left[1 - \frac{\rho l^2}{M} \cdot \frac{1}{ho} \left\{ \frac{\pi^2}{6} \left(\frac{L}{a} \right)^2 m^2 - 1 \right\} \right]^{\frac{1}{2}}$$

m=odd ....(7.2)

This expression corresponds to the natural circular frequency $W_{om} = W_{mo}$ given by equation (5.20). However, the frequency of this expression decreases, as the cylinder number N= 8/L increases (**Fig. 7.5**), in the same manner as the results of discrete parameter model. For n→∞, this concedes with that of equation (5.20).

7.3 Three Parallel Circular Cylindrical Tubes Immersed in a Liquid

From the study of **Chapter VI**, it is seen that:

(a) Added mass coefficients for each cylinder, when boundary conditions are same, are identical.

(b) The equations of frequency (6.26), is of 6 modes, but from equation (6.20), it is evident that equations for tubes 'I' & '3' are identical whereas 'I' are different due effect of fluid coupling by 'I and '3'. It is also very clear that if tubes are large, the equations of motion, for tubes near the shells are same and for intermediate tubes, it is also identical to the equation of tube '2'. This can be easily seen by adopting analytical approach for four cylinders and or more cylinders.

(c) It is already assumed that tubes are same diameter, and isotropic, thus the efficiency of system can be higher only when Gap/radius ratio is higher and also number tubes in the system is large.

(d) As already studied from para (7.1) (d), the coupling between rods depends on the radius ratio (here it is 1) gap-radius ratio = G/R and mass ratio 'B' (in this case it is 1). Thus fluid crippling between rods depends on the Gap- radius ratio (G/R). Thus, if G/R is large the tubes will respond in dependently.

(e) Similarly, solutions for frequency can also be derived for three cylinders for the forced vibration and considering approximate solutions.

CHAPTER-VIII

CONCLUSIONS

8. Conclusions

The analytical approach for attending the vibration problem of Heat Exchanger tubes are more combursum. Thus many studies are also made experimentally to verify this results obtained by this approach. However, efforts should be made to develop the analytical method for finding out the frequency response of tubes, when numbers are large. In this study author tried to derive the frequency equation for three cylinders, which resulted that still efforts can be made to derive the equations of frequency for smaller or large number of cylinders, but it would be identical to the equations of motion (6.13), (6.20) and frequency equation (6.26) if boundary conditions are choosen as under:

(i) Radii of all cylinders are same as 'R', thus Ratio of radii ($R_{j+1}/R_j=1$)

(ii) Material of all cylinders/tubes are same thus mass ratio ($\beta=1$).

It is concluded that if the G/R ratio is kept large, cylinders will respond independently, the coupling effect will reduce the same time, efficiency of the heat exchanger will also reduce. The efficiency of the Heat transmission from Heat exchanger-tube Bundle will be represented as:

$$\eta_H = F\left(\frac{G}{R}, V, \right) \quad \ldots (8.1)$$

Where η_H = efficiency off Heat transmission of Heat Exchanger.

G = Gap between Tubes,

R = radius of tubes &

V = flow velocity of fluid

We have already seen from the study of two cylinders Ref. (5)

(1) For $\dfrac{G}{R} = 0.2$, coupling is strong (r = 2 R or d)

(2) For $\dfrac{G}{R} = 0.2$, coupling is weak (r = 4 R or 2d)

This also satisfies the study done by B.W. Bearman [4], S.S. Chen, [23] also tried to present in his paper about relation between added mass coefficients to the pitch-diameter ratio when tubes are large in number & given a pattern of lower & upper bound of effective added mass coefficients. Moreover, if we are able to develop the relations of the η_H (efficiency of Heat transmission of Heat Exchanger) with Gap-radius, ratio, fluid flow velocity and geometries of tubes, our international problem for getting maximum efficiency from Heat exchanger without frequent failure of tube are solved. Thus, author leaves this problem open to all to work on experimental as well as analytical approach to find out the above relations and future study of author would also be continued on this topic.

CHAPTER-IX

REFERENCES AND BIBLIOGRAPHY

9. References & Bibliography

[1]. Masami Masubuchi, "Dynamic Response and Control of Multipass heat Exchangers", transaction of the ASME Journal of Basic Engineering, March, 1960, Pages 51 to 65.

[2]. Herman Thal-Larsen, "Dynamics of heat Exchangers and Their Models", Transactions of the ASME Journal of Basic Engineering, June 1960, Pages 489 to 504.

[3]. A. Protos, V.W. Goldsmichdt, G.H. Toebes "Hydro-elastic Forces on Bluff Cylinders", Transactions of the ASME, Journal or Basic Engineering, September, 1968, Pages 378 to 386.

[4]. P.W. BEARMAN & A.J. WADCOCK, "The Interaction Between a pair of circular Cylinders normal to a stream", Journal of Fluid Mechanics (1973) Vol. 61, part 3, pages 499 to 511.

[5]. S.S. CHEN, "Dynamic Responses of Two Parallel Circular Cylinders in a Liquid", Transactions of the ASME, Journal of Pressure Vessel Technology, May, 1975, Pages 78 to 83.

[6]. Y. Shinohara, T. Shimogo, "Vibrations of square and Hexagonal cylinders in a Liquid", Transactions of the ASME, Journal of Pressure Vessel Technology, Aug. 1981, Vol. 103, Pages 233 to 239 .

[7]. S. S. Chen, J. A. JENDRAZEJCZYK, "Flow Velocity Dependence of Damping in Tube Arrays subjected to Liquid Cross Flow", transactions of the ASME, Journal of Pressure Vessel Technology, May, 1981, Vol. 103, Pages 130 to 135.

[8]. S. J. Brown, "A survey studies into the Hydrodynamic Response of Fluid coupled Circular Cylinders" Transactions of the ASME, Journal of the Pressure Vessel Technology, Feb. 1982, Vol. 104, Pages 2to 19.

[9]. D. S. Weaver, D. Koroyannakis, "The Cross-Flow Response of the Tube Array in Water-A comparison with the same Array in Air" Transactions of the ASME, Journal of Pressure Vessel Technology Aug. 1982, Vol. 104, Pages 139 to 146.

[10]. H. Tanaka, S. Takahara, K. Ohta, "Flow Induced Vibration of Tube Arrays with various Pitch to Diameter Ratios" Transactions of the ASME, Journal of Pressure Vessel Technology, Aug. 1982, vol. 104, Pages 168 to 174.

[11]. K. Ohta, K. Kagawa, H. Tanaka, S. Takahara, "study on the Fluid elastic Vibration of Tube Arrays using Model Analysis Technique", Transactions of the ASME, Journal of Pressure Vessel Technology, Feb. 1984, Vol. 106, Pages 17 to 24.

[12]. S. S. Chen, J.A. Jendrazejczyk, M.W. Wambsganss, "Dynamics of Tubes in Fluid with Tube Baffle interaction" Transactions of the ASME? Journal of Pressure Vessel Technology, Feb, 1985, Vol. 107, Pages 7 to 17.

[13]. G.J. Rae & B.C. Murray, "Flow Induced Acoustic Resonances in Heat Exchangers", International Conference on Flow Induced Vibrations, Bowness on Windermere, England, 12-14 May, 1987 Pages 221 to 231.

[14]. H.G.D. Goydev, "Two phase Buffeting of Heat Exchanger Tube", International Conference on Flow Induced Vibrations, Bowness on Windermere, England, 12-14 May 1987, Pages 211 to 219.

[15]. Fumio Hara, "Vibration of A single Row of Circular Cylinders subjected to Two-Phase Bubble Cross Flow" International Conference on Folo Induced Vibrations Bowness on Windermere, England: 12-14 May 1987, Pages 203 to 210.

[16]. P.K. Stanslay, P.A. Smith, R. Penoyre, "Flow Around Multiple Cylinder by the Vortex Method" International Conference on Flow Induced Vibrations Bowness on Windermere, England: 12th to 14th May, 1987 Pages 41 to 50.

[17]. D.S. Weaver, J.A. Fitzpatrick, "A Review of Flow Induced Vibrations in Heat Exchangers", International Conference on Flow Induced Vibrations", Bowness-on-Windermere, England: 12-14 May, 1987, Pages 1 to 17.

[18]. J.P. Giesing, "Nonlinear Interaction of Two Lifting Bodies in Arbitrary unsteady Motion", Transactions of the ASME, Journal of Basic Engineering, Sept. 1968, Pages 387 to 394.

[19]. S.S. Chen, "Vibrations of a Row of Circular Cylinders in a Liquid" Transactions of the ASME, Journal of Engineering for Industry, Nov. 1975, Pages 1212 to 1218.

[20]. R.D. Blevins, "Vibration of a Loosely Held Tube" Transactions of the ASME, Journal of Engineering for Industry Nov. 1975, pages 1301 to 1304.

[21]. B.T. Lubin, K.H. Haslinger, A Puri, J. Goldberg, "Frequency Response of a Tube Bundle in Water", Transactions of the ASME, Journal of Fluid Engineering June 1977, pages 416 to 418.

[22]. R.D. Blevins, "Fluid Elastic Whirling of Tube Rows and Tube Arrays" Transactions of the ASME, Journal of Fluids Engineering, Sep. 1977, pages 457 to 461.

[23]. S. S. Chen, "Dynamics of Heat Exchanger Tube Banks", Transactions of the ASME, Journal of Fluids Engineering Sept. 1977, Pages 462 to 469.

[24]. M.K. Auyang, "Turbulent Buffeting of a Multispan Tube Bundle", Transactions of the ASME, Journal of Vibration, Acoustics, Stress and Reliability in design, April 1986, Vol. 103, pages 150 to 154.

[25]. Fraciess. Tse, Ivan E. Morso, Rolland. T. Hinkla Mechanical Vibrations, Theory and Applications First Indian Edition 1983.

[26]. Kewal Pujora, "Vibrations & noise for Engineers" Dhanpat Rai & Sons New Delhi, Second Edition, 1977.

[27]. Bonilla C.F., Nuclear Engineering, Mc Graw Hill.

[28]. Thomas J. Connolly "Foundation of Nuclear Engineering" J. Willey & Sons.

[29]. E.C. Ting and A. Hosseinipour, Journal of Sound and Vibration, Vol. 88, May-June, 1983, Page 289-298.

[30]. D. Jha and K.N. Gupta, "Flow-Excited Vibration of Nonlinear Panels", paper no. J5, proc pp 593-603, published in proceedings of the International Conference on Flow-Induced Vibrations, Bowness-on-Windermere, May 12-14, 1987, England organized and sponsored by the British Hydromechanics Research Association, England and Co-sponsored by the U.K. Atomic Energy Authority Wind-scale Laboratories, The British

CHAPTER-X

TABLES, APPENDICES & GRAPHS

10.

Tables, Appendices & Graphs

Table-1: Natural frequency to Different Tube Diameters & Thickness

S. No.	Diameter of tube	Thickness of tube	$1 = \frac{\pi}{64}\left(D_0^4 - D_1^4\right)$	Natural Frequency 'w_n' $= \frac{1}{2\pi}\frac{(\pi)^2}{L}\sqrt{\frac{EIg}{m}}$	
					Cycle/sec
1.	26.9 mm	2.35	1.0420	1.41	6.1259
	(20NB)	2.65	1.2261	1.58	6.2774
		3.25	1.6358	1.90	6.6120
2.	33.7 mm	2.65	2.2206	2.010	7.4900
	(25NB)	3.25	2.9162	2.440	7.7904
		4.05	3.9798	2.970	8.2448
3.	42.4 mm	2.65	4.3557	2.580	9.2590
	(32NB)	3.25	5.6391	3.140	9.5496
		4.05	7.5473	3.840	9.9902
4.	48.3 mm	2.90	9.0349	3.250	11.8813
	(40NB)	3.25	10.3862	3.610	12.0870
		4.05	13.7126	4.430	12.5372
5.	60.3 mm	2.90	16.9137	40110	14.4558
	(50NB)	3.65	22.2423	5.100	14.8816
		4.50	28.8087	6.170	15.3980

Sample Calculations: $w_n = \frac{1}{2\pi}\frac{(\pi)^2}{L}\sqrt{\frac{EIg}{m}}$ cycle/sec

Where (i) $E = 2.1 \times 10^6$ kg/cm^2

(ii) $g = 981$ cm/sec

(iii) Length of tube $= 100$ cm

$$wn = \frac{1}{2} \times \frac{3.14}{10^4}\sqrt{2.1 \times 981 \times 10^6} \times \sqrt{\frac{1}{m}}$$

$$= \frac{45.3883 \times 3.14}{2 \times 10}\sqrt{\frac{1}{m}} = 7.126 \times \sqrt{\frac{1}{m}}$$

Table-2: Mass Coefficient versus Gap Ratios of Tubes

S. No.	G/R or $r = \left(2+\dfrac{G}{R}\right)R$	h	$h_1=h_2$	$\dfrac{r^2-4R^2}{R^2}$	$\dfrac{r^2(r^2-4R^2)}{R4}$	$e^{-(h+h_1)}$	K=1 $e^{-(h+h_1)}$	Sinh (h)	$\dfrac{2(h+h_1)}{2e}$	K=1 -4h he	Sinh (2h)	$\mu_1=\mu_2$
1.	$0, r=2.005_R$	0.1414	0.1414	0.0200	0.0805	0.7537	0.1066	0.1418	1.1360	0.0803	0.2865	1.3735
2.	$0, r=2.01_R$	0.1993	0.1999	0.0401	0.1620	0.6713	0.1338	0.2006	0.9012	0.0898	0.4092	1.3600
3.	$0, r=2.05_R$	0.4463	0.4463	0.2025	0.8510	0.4096	0.1828	0.4612	0.3355	0.0749	1.0159	1.2734
4.	$0.1, r=2.10_R$	0.6298	0.6298	0.4100	1.8081	0.2838	0.1787	0.6722	0.1610	0.0507	1.6201	1.2132
5.	$0.2, r=2.2_R$	0.8871	0.8871	0.8400	4.0656	0.1696	0.1505	1.0081	0.0575	0.0255	2.8629	1.1596
6.	$0.4, r=2.4_R$	1.2447	1.2447	1.7600	10.1376	0.0830	0.1033	1.5919	0.0138	0.0086	5.9855	1.0968
7.	$0.6, r=2.6_R$	1.5129	1.5129	2.7600	18.6576	0.0485	0.0734	2.1598	0.0047	0.0036	10.2807	1.0690
8.	$0.8, r=2.8_R$	1.7340	1.7340	3.8400	30.1056	0.0312	0.0541	2.7433	0.0019	0.0017	16.0206	1.0384
9.	$1.0, r=3.0_R$	1.9248	1.9248	5.0000	45.0000	0.0213	0.0410	3.3539	0.0009	0.0009	23.4764	1.0250
10.	$1.5, r=3.5_R$	2.3176	2.3176	8.2500	101.0625	0.0097	0.0225	5.0263	0.0002	0.0002	51.5194	1.0050
11.	$2.0, r=4.0_R$	2.6339	2.6339	12.0000	192.0000	0.0052	0.0136	6.9280	0.00005	0.0001	96.9917	1.0000
12.	$2.5, r=4.5_R$	2.9011	2.9011	16.2500	329.0625	0.0030	0.0088	9.0695	0.00002	0.00002	165.512	1.0000
13.	$3.0, r=5.0_R$	3.1336	3.1336	21.0000	525.0000	0.0019	0.0059	11.4564	0.000007	0.00001	263.4980	1.000

(* See Appendix-1 for sample calculations)

Vibration of Heat Exchanger Tubes

Table-3: Frequency Ratios versus Mass Ratios

S. No.	Dia of rod d	$Mj = \frac{\rho \pi d^2 l}{4000}$ in kg.	$Mj = \frac{\pi d l}{1000}$ in kg.	$Mj/mj = \beta j$	$\left(W_m/\overline{W_{in}}\right) = 1/\sqrt{1+\mu_j\left(\frac{Mj}{mj}\right)}$			
					G:R = 0.005 or μ = 1.3725	G:R = 0.01 or μ = 1.300	G:R = 0.05 or μ = 1.2734	G:R = 0.10 or μ = 1.2132
1)	20 mm	2.4660	0.6283	0.2548	0.8607	0.8618	0.8689	0.8740
2)	25 mm	3.5831	0.7854	0.2038	0.8839	0.8849	0.8910	0.8954
3)	32 mm	6.3130	1.0053	0.1592	0.9059	0.9067	0.9118	0.9155
4)	40 mm	9.8640	1.2566	0.1274	0.9225	0.9232	0.9276	0.9307
5)	50 mm	15.4125	1.5708	0.1019	0.9366	0.9372	0.9408	0.9434

Sample calculations

Applying natural frequency in vacuo for 'n' the node

$$\left(\frac{W_{jn}}{\overline{W_{jn}}}\right) = 1/\sqrt{1+\mu_j\beta_j} \text{ s where } \beta_j = \frac{Mj}{m_j}$$

Where Wjn = Frequency of rod close to a rigid rod.
$\overline{W}_{jn}$ = Frequency of nth mode in vacuo
μ_A = Added mass coefficient

$\beta_j = \left(\frac{M_j}{m_j}\right)$ = mass ratio

l = length of rod = 100 cm
M_j = displaced mass of fluid = $\pi.d.l \times 10^{-3}$ kg
m_j = mass per unit length of rod
$= \frac{\rho \pi d^2 l}{4} \times 10^{-3 kg (0.785 \times 14.1006 \times l/2 \cdot l)}$
$\rho = 7.85$ gm/cm^3 = 0.6165 d^2

Table-4: Different Parameters of Square Cylinders

S. No.	Square Cylinder (l^2)	$M = \left(\dfrac{l^2 \rho_c \cdot 100}{1000}\right) kg$	$(\rho l^2/M)$	W_{om}/W_o				
				N=1 $\{=4.2899\}$	N=2 $\{=1.8225\}$	N=3 $\{=1.3655\}$	N=10 $\{=1.0987\}$	N=30 $\{=1.003\}$
1)	20 mm x 20 mm	0.8	3.925	0.0530	0.1075	0.1635	0.6038	2.6166
				0.9027	0.9144	0.9042	0.7754	0.5251
2)	25 mm x 25 mm	1.0	4.9063	0.0831	0.1692	0.2582	0.9812	4.9063
				0.8585	0.8742	0.8598	0.6937	0.4108
3)	32 mm x 30 mm	1.2	5.8875	0.1201	0.2453	0.3758	1.4718	8.8312
				0.8124	0.8313	0.8129	0.6181	0.3184
4)	40 mm x 40 mm	1.6	7.8500	0.2150	0.4422	0.6826	2.8545	31.400
				0.7212	0.7441	0.7194	0.4917	0.1753
5)	50 mm x 50 mm	2.0	9.8125	0.3383	0.7009	1.0902	4.9062	-
				0.6387	0.6626	0.6339	0.3956	-

Sample Calculations:

Applying Energy equation for Square Cylinders

Natural circular frequency, when m = 0, n = 0 or m = 0, n = 0 or m = n

Then $W_{om} = W_{mo} = W_{mm} = W_o \left[1 + \dfrac{\rho l^2}{M}\left(\dfrac{1}{ho}\right)\left(\dfrac{L}{a}\right)^2 \left(\dfrac{\pi}{3}m^2 + 1\right)\right]^{-\frac{1}{2}}$

Where m = 1, 2, ...

Where $W_o = \sqrt{K/M}$ = natural frequency of this cylinder without coupling

h_o = gap with in equilibrium
l = length of one side of cylinder
M = mass of cylinders per unit length
N = number of cylinders in each row (a/L)

When a = 1500 mm, l = 20 mm

$N = \dfrac{a}{L} = 1, a = L$ ....ho = L-1 = 1500 − 20 = 1480

$N = 2, L = a/2$ ......ho = L-1 = $\dfrac{a}{2} - \ell = 730$

$N = 3, L = a/3$ ......ho = L-1 = $\dfrac{a}{3} - 1 = 480$

$N = 10, L = a/10$ ......ho = $\dfrac{a}{10} - 1 = 130$

$N = 30, L = a/30$ ......ho = $\dfrac{a}{30} - 1 = 30$

Table-5: Different Parameters for Various Sizes of Tube Diameters

S. No.	Tube (Di)	dia D_o	Thickness t	Ms	$Mf = \dfrac{\pi D_o^2}{4} \cdot 1 \cdot \xi$ $= 78.54\ Do^2 \times 10^{-3}$	μ when G/R=1	$M_h = \mu\ Mf$	Mfi	$f_n\ f_{cr} = \left[1 + \left(\dfrac{\mu \Delta f + Mf}{Ms}\right)\right]$
(1)	20 mm	24.7	2.35	1.410	0.4792	1.05	0.5032	0.10	0.8369
		25.3	2.65	1.580	0.5027	1.05	0.5278	0.10	0.8460
		26.5	3.25	1.900	0.5515	1.05	0.5791	0.10	0.8583
(2)	25 mm	30.3	2.65	2.010	0.7211	1.05	0.7572	0.1563	0.8292
		31.5	3.25	2.440	0.7793	1.05	0.8183	0.1563	0.8453
		33.10	4.05	2.970	0.8605	1.05	0.9035	0.1563	0.8585
(3)	32 mm	37.3	2.65	2.580	1.0927	1.05	1.1473	0.2560	0.8048
		38.5	3.25	3.140	1.1642	1.05	1.2224	0.2560	0.8246
		40.1	4.05	3.840	1.2629	1.05	1.3260	0.2560	0.8416
(4)	40 mm	54.8	2.90	3.250	1.6475	1.05	1.7299	0.4000	0.7772
		46.5	3.25	3.610	1.6982	1.05	1.7831	0.4000	0.7894
		48.1	4.05	4.430	1.8171	1.05	1.9080	0.4000	0.8108
(5)	50 mm	55.8	2.90	4.110	2.4455	1.05	2.5678	0.6250	0.7502
		57.3	3.65	5.100	2.5787	1.05	2.7076	0.6250	0.7777
		59.0	4.50	6.170	2.7340	1.05	2.8707	0.6250	0.7990

* (See Appendix – II for sample calculations)

APPENDIX –I

Sample Calculations: Mass Coefficient with Different Gap Ratios

$$\mu_1 = 1 + \frac{r^2 - 4R^2}{R^2} \sum_{K=1}^{\infty} \cdot \frac{Ke^{-k(h_1+h_2)}}{\text{Sinh}(kh)}$$

(i) when $\gamma = \underline{2.005}$ or $\underline{G/R = 0.005}$

$$\mu_1 = 1 + 0.02 \begin{Bmatrix} \dfrac{0.7537}{0.1418} + \dfrac{1.136}{0.2865} + \dfrac{1.2843}{0.4370} + \dfrac{1.2906}{0.5962} + \\ \dfrac{1.2160}{0.7674} + \dfrac{1.0996}{0.9539} + \dfrac{0.9669}{1.1595} + \dfrac{0.8338}{1.338} \end{Bmatrix}$$

$$= 1.3735$$

(ii) When $\underline{r = 2.01}$ or $\underline{G/R = 0.01}$,

$$\mu_1 = 1 + 0.0405 \begin{Bmatrix} \dfrac{0.6713}{0.2006} + \dfrac{0.9012}{0.4092} + \dfrac{0.9041}{0.6363} + \\ \dfrac{0.8121}{0.8844} + \dfrac{0.6814}{1.1698} + \dfrac{0.5489}{1.5019} + \dfrac{0.4299}{1.8938} \end{Bmatrix}$$

$$= 1.3609$$

(iii) When $\underline{r = 2.05}$ or $\underline{G/R = 0.05}$,

$$\mu_1 = 1 + 0.2025 \left\{ \dfrac{0.4096}{0.4613} + \dfrac{0.3353}{1.0159} + 0 \dfrac{0.2061}{1.7764} + \dfrac{0.1126}{2.896} + \right\}$$

$$= 1.2754$$

(iv) When $\underline{r = 2.2}$ or $\underline{G/R = 0.2}$,

$$\mu_1 = 1 + 0.84 \left\{ \dfrac{0.1696}{1.0081} + \dfrac{0.0576}{2.8630} + \dfrac{0.0146}{7.1226} + \right\}$$

$$= 1.1596$$

(v) When $\underline{r = 2.4}$ or $\underline{G/R = 0.4}$,

$$\mu_1 = 1 + 1.76 \left\{ \dfrac{0.0830}{1.5919} + \dfrac{0.0138}{5.9885} + \dfrac{0.0018}{20.9132} \right\}$$

$$= 1.0968$$

APPENDIX –II

Sample calculations:

Frequency Response of a Tube Bundle Emerged in Water

Virtual Mass = $M_s + M_{fi} + M_h$ (1)

M_s = mass of solid

M_{fi} = internal fluid mass

M_h = equivalent hydrodynamic mass/unit length

M_h = $C_m M_f$ ----------------   (2)

M_f = mass per unit length displaced fluid

C_m = hydrodynamic mass coefficient

$$\left(\frac{fn}{fo}\right) = 1 / \left[1 + \left(\frac{C_m M_f + M_{fi}}{M_s}\right)\right]^{\frac{1}{2}} \quad \ldots\ldots\ldots(3)$$

Graph Sheet No. 1

Natural frequency Versus Different Tube Diameters & Thickness

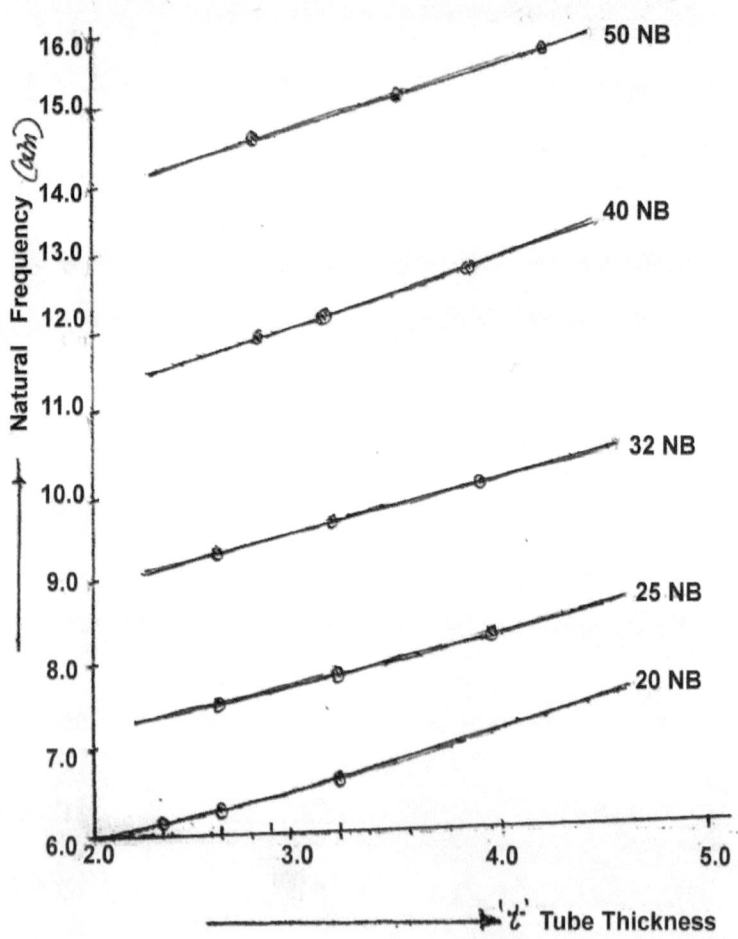

Graph Sheet No. 2

Mass Coefficient vs. Gap Ratios of Tubes

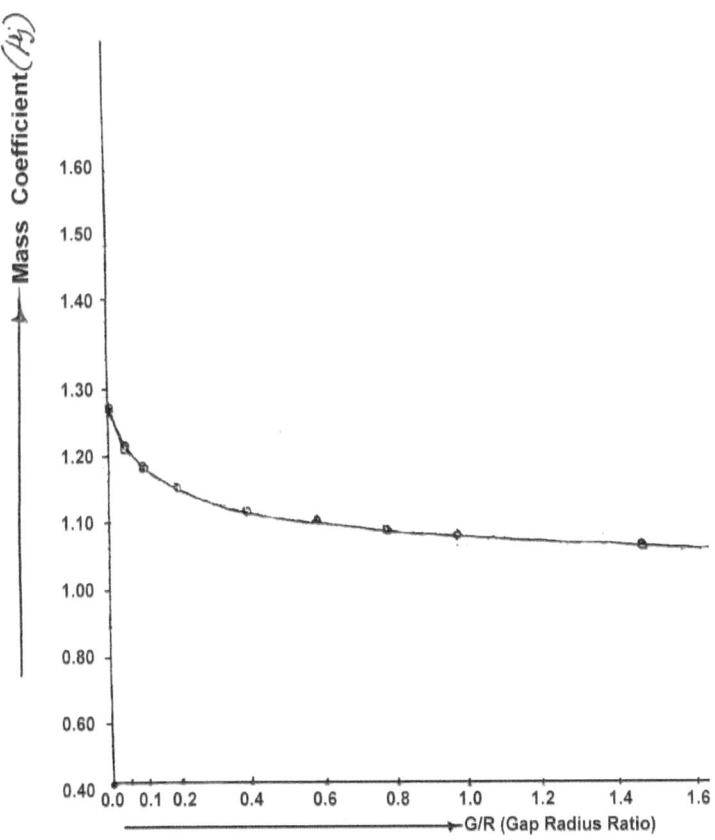

Graph Sheet No. 3

Frequency Ratio vs. Mass Ratios

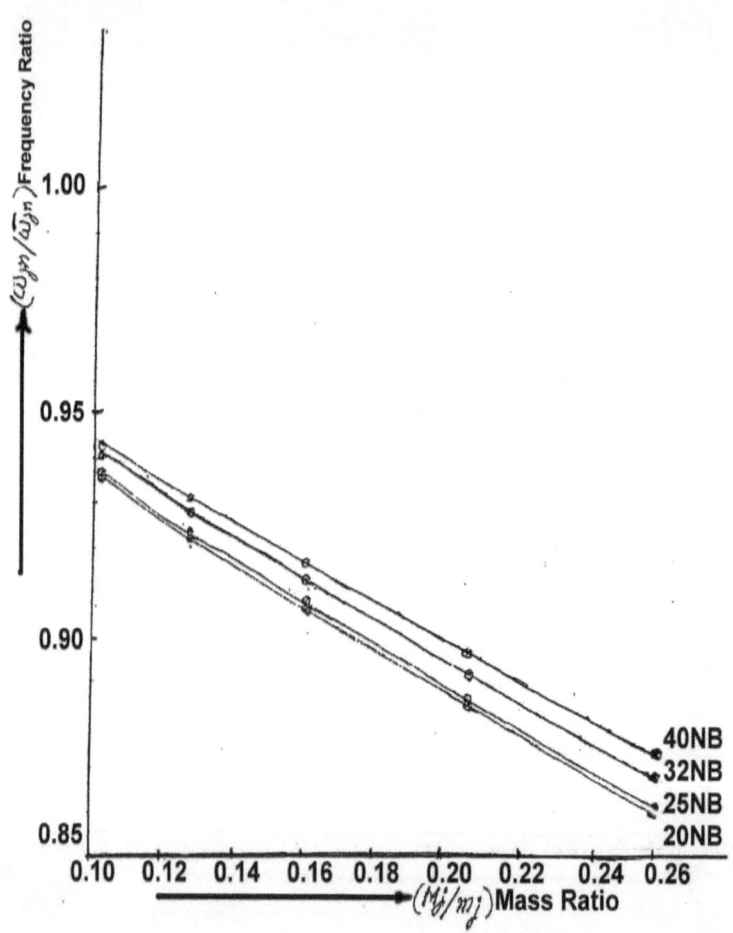

Graph Sheet No. 4

Frequency Ratio vs. Mass /Gap Ratio

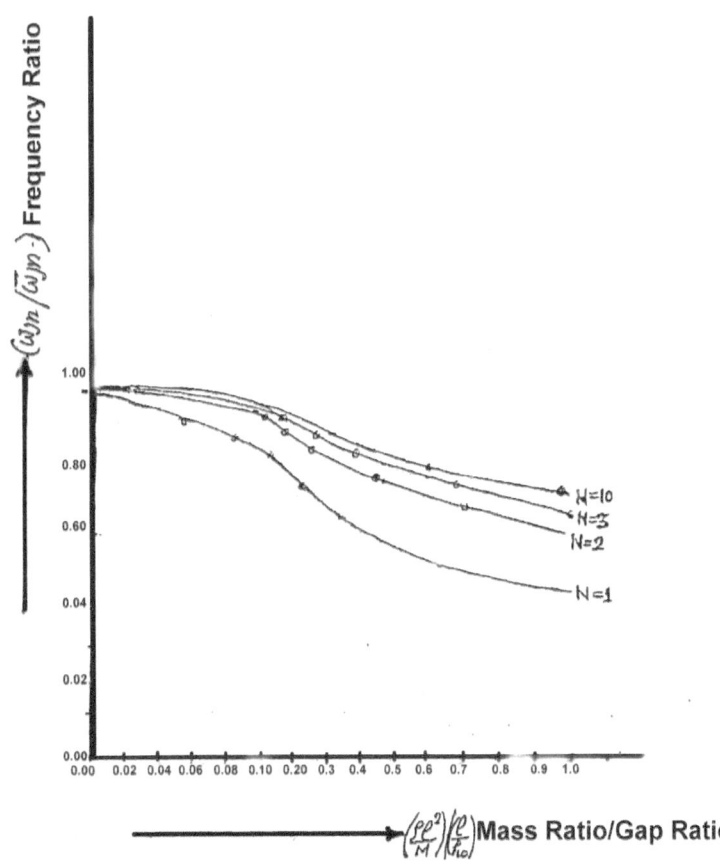

Graph Sheet No. 5

Frequency Ratio vs. Tube Diameters with Different Thicknesses

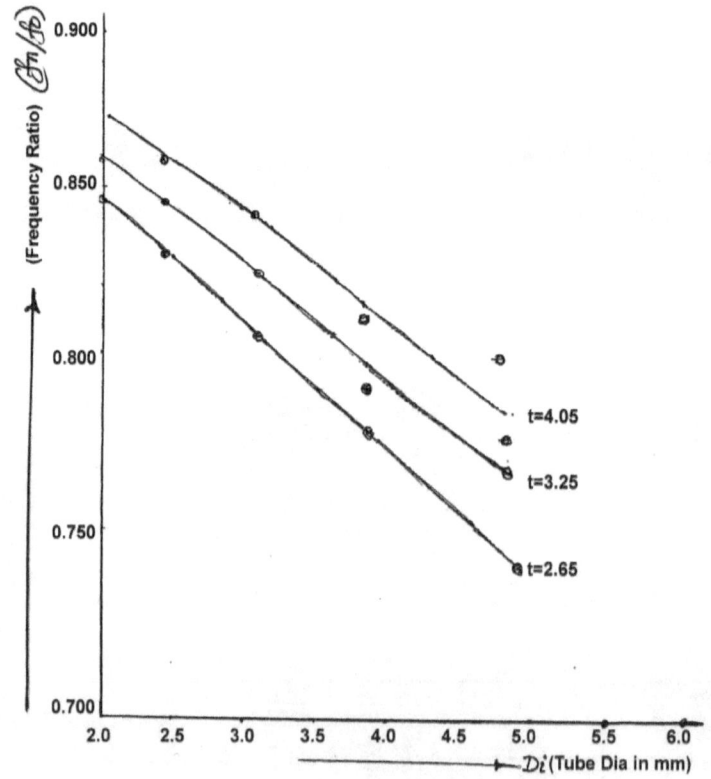

The Vibration of Heat Exchanger Tubes
Published on: Dec 07, 2019
First Edition
ISBN: 978-1-79479-184-8
Price: US Dollar $10.49 (Rs.750/=)

Publisher:
Lulu Press, Inc.
627 Davis Drive, Suite 300, Morrisville, NC 27560, USA
www.lulu.com

Copyright © Lulu Press, Inc. All Rights Reserved.

www.ingramcontent.com/pod-product-compliance
Lightning Source LLC
Chambersburg PA
CBHW021434210526
45463CB00002B/510